PAINTING THE WHITE HOUSE GREEN

Rationalizing Environmental Policy Inside the Executive Office of the President

EDITED BY
Randall Lutter
Jason F. Shogren

Resources for the Future
Washington, DC, USA

An RFF Press book
Published by Resources for the Future
1616 P Street NW
Washington, DC 20036–1400
USA
www.rffpress.org

Library of Congress Cataloging-in-Publication Data

Painting the White House green : rationalizing environmental policy inside the executive office of the president / edited by Randall Lutter and Jason F. Shogren.
   p.   cm.
Includes bibliographical references and index.
ISBN 1-891853-73-2 (hardcover : alk. paper) -- ISBN 1-891853-72-4 (pbk. : alk. paper)
1. Environmental policy--United States. I. Lutter, Randall W. II. Shogren, Jason F.
GE180.P35 2004
333.72'0973--dc22   2004010086

f e d c b a

The paper in this book is recycled and meets the guidelines for permanence and durability of the Committee on Production Guidelines for Book Longevity of the Council on Library Resources.

This book was typeset in Giovanni and Myriad by Peter Lindeman. Interior design by Naylor Design, Inc. It was copyedited by Joyce Bond. The cover was designed by Rosenbohm Graphic Design.

ISBN 1-891853-72-4 (paper)   ISBN 1-891853-73-2 (cloth)

About Resources for the Future *and* RFF Press

Resources for the Future (RFF) improves environmental and natural resource policymaking worldwide through independent social science research of the highest caliber. Founded in 1952, RFF pioneered the application of economics as a tool for developing more effective policy about the use and conservation of natural resources. Its scholars continue to employ social science methods to analyze critical issues concerning pollution control, energy policy, land and water use, hazardous waste, climate change, biodiversity, and the environmental challenges of developing countries.

RFF Press supports the mission of RFF by publishing book-length works that present a broad range of approaches to the study of natural resources and the environment. Its authors and editors include RFF staff, researchers from the larger academic and policy communities, and journalists. Audiences for publications by RFF Press include all of the participants in the policymaking process—scholars, the media, advocacy groups, NGOs, professionals in business and government, and the public.

CONTENTS

FOREWORD

Janet Yellen

Established by the Employment Act of 1946, the Council of Economic Advisers (CEA), an economic consulting group within the White House, is one of the great creations of the American government. CEA's job is to provide unbiased, objective analysis and independent advice to the president on every policy under consideration by the White House with a significant economic component, which is almost everything these days, including environmental regulation. CEA is one of the smallest agencies in Washington, consisting of three members; roughly ten senior economists, mainly one-year rotators from academia; and a small support staff. While others have budgets of billions, CEA subsists on a budget considerably smaller than that of a medium-size department of economics and about one-hundredth the amount that the Federal Reserve spends annually in support of monetary policy. Nevertheless, CEA consistently attracts to its ranks outstanding economists with a passion for public policy, including the authors of this volume. CEA economists almost always have a seat at the table and, for more than half a century, have made their presence felt.

This fascinating volume recounts the experiences of eight persons who served as senior economists at CEA with responsibility for environmental and natural resource issues during the 1990s and early 2000s, in the administrations of both George Bushes and Bill Clinton. It was my pleasure and honor to have worked with four of them during my term as CEA chair, from 1997 to 1999. The authors were personally involved in key environmental policy decisions: the development of new standards for ozone and particulate emissions under the Clean Air Act, the design of proposed federal legislation to support electricity restructuring, an initiative to protect forests in developing countries, and the formulation of policies to combat global warming, including the negotiation of the 1997 Kyoto Protocol.

Readers interested in policymaking in the executive branch will find detailed accounts of the White House policy process and frank and revealing portraits of the dynamics, tensions, and power plays involved in environmental policymaking. The authors provide vivid descriptions of lively White House debates on clean air, climate change, and electricity deregulation that pitted economists at CEA, the Office of Management and Budget, and often the Treasury Department against political advisers in the White House and officials at EPA and various other cabinet agencies.

This book should be required reading for those interested in understanding the role that economic analysis plays in policymaking generally and environmental policy in particular. According to its mandate and tradition, CEA's job in the White House is to provide analysis and input representing the best professional thinking on the issues under consideration, advocating policies that promote the broad national interest. Rather than represent the well-informed, well-organized special interests, CEA's job is to speak for the large numbers of people who are either unaware of or limited in affecting environmental policies. CEA is the champion of efficiency in the design of policies; the agency routinely seeks to identify and quantify costs, benefits, risks, and tradeoffs to promote rational policy choices. CEA's normal role is to see that bad ideas generated elsewhere in the government do not turn into policy, and that those policies that are adopted improve societal welfare and achieve maximum bang per buck. A common CEA burden is to be the bearer of bad news—to tell the president and his advisers that environmental cleanups are costly and that new technologies are unlikely to provide a free lunch, a role that once provoked an annoyed Bill Clinton to refer to his economic advisers as "lemon suckers."

In contrast to macroeconomic policy, where disagreement among economists is common, in the arena of microeconomic policy, agreement among economists is the norm, so the advice that presidents receive from CEA is similar in both Democratic and Republican administrations. On environmental issues, economists agree that regulations are needed to protect the environment: the issues are not whether but how to regulate, and just how stringent environmental goals and standards should be. Balancing costs and benefits, CEA staffers commonly advocate a middle course, engendering hostility from both environmentalists, who too often show little interest in the costs of combating pollution, and businesses, which too often seem unconcerned about the environment.

The essays in this volume describe the efforts of eight CEA staffers to inject economic rationality into environmental policymaking. In their attempts to shape environmental policy in the national interest, the staffers view their records of success as decidedly mixed. Many are sharply critical of environmental policy outcomes and frustrated by the frequency with which politics and emotion trump economics and rational argument. In the case of the 1997 National Ambient Air Quality Standards, Randall Lutter laments EPA's refusal to weigh ozone's health benefits in preventing skin cancer and cataracts along with its adverse effects on respiratory health, noting that this failure to do so left the EPA rule vulnerable to later court challenges. He is also sharply critical of CEA's estimates of the likely costs of its new regulation. Stephen Polasky describes how environmental constituencies fought to raise the quantitative requirements for renewable energy in the Clinton administration's electricity restructuring bill above reasonable levels, even though, as he explains, the victory was purely symbolic. Michael Toman is critical of the Clinton administration's Climate Change Initiative, particularly its stance toward the Kyoto Protocol; here, CEA's pleas for cost-effective targets and timetables at Kyoto went down in defeat, as did its advocacy of a safety valve to guard against the heavy costs that might result if the Kyoto emissions requirements turned out to be too stringent. CEA also favored transparent and well-targeted near-term policies, such as a carbon tax, to at least begin the long process of addressing this potent environmental threat. Economists view emissions taxes as straightforward and effective means to penalize polluters for the damage they inflict on the environment. As Toman notes, however, taxes are so unpopular as to be politically taboo and hence were off the table as a strategy to address global warming. Jason Shogren blames politicians in Washington for enacting environmental regulations that often compound rather than alleviate the problems faced by western states. In their perceptive essays, Jason Shogren and Ray Squitieri offer some hypotheses to explain why economists are so frequently on the losing side of policy debates. The authors appropriately emphasize the need for economists to refine their skills in communicating with noneconomists and improve their understanding of the inherently political processes that dominate decision making in the White House.

Despite the setbacks documented by the staffers, I hope that readers of this volume will appreciate the overall positive influence that CEA has had on White House decision making and that economic analysis has had on the evolution of environmental policy in the United States. Joseph Aldy and Jonathan Wiener document some of CEA's successes. On policy toward global warming, CEA had a strong influence on policy in the post-Kyoto period. The decision by the Clinton administration to present an economic analysis of the Kyoto Protocol and its proposed climate policies to Congress, along with CEA's key role in preparing and defending that analysis, enhanced CEA's influence and constrained later policy and diplomatic choices. The Clinton administration remained persistent in its advocacy of cost-effective international arrangements to meet the Kyoto Protocol goals and worked intensively with developing countries to devise beneficial ways for them to undertake climate change commitments. Jonathan Wiener describes the successful effort spearheaded by the first Bush administration economists to establish a federal fund to compensate landowners in developing countries for preserving forests, resulting in the Forests for the Future Initiative launched by the United States at the 1992 Earth Summit in Rio. William Pizer examines the shift in climate change policy in the second Bush administration, acknowledging how economic reasoning competes with the many noneconomic concerns that ultimately drive policy.

In my experience, CEA racked up few total wins but achieved successes large and small in nudging policy in more rational directions. In America, the air is cleaner, the water is purer, and the regulatory approaches that are being used to attain these objectives are typically more market-friendly and cost-effective now than in decades past. A long-standing executive order mandates that federal agencies must prepare assessments of the costs and benefits of new federal regulations and choose cost-effective regulatory strategies, even though, as Randall Lutter documents, the practice sometimes falls short of the ideal. Incentives and market mechanisms, rather than command and control style approaches, are now more common in environmental regulation. Tradable pollution permits have been used with success by EPA to reduce emissions of SO_2 by electric ultilities and in the phasedown of lead in gasoline; tradable emissions permits figure prominently in the methodologies being considered to curb global warming. The broad acceptance of the merits of such mechanisms by both environmentalists and businesses represents a public policy victory for economics.

Painting the White House Green is an invaluable resource for economists and noneconomists alike who seek a deeper understanding of key environmental issues facing the country and the process by which environmental policy is made. Through firsthand accounts of the debates that took place behind the closed doors of the White House, the authors of this insightful and readable volume illustrate the reasoning that economists bring to the table and the sometimes subtle ways in which their participation works to improve environmental regulation.

JANET YELLEN

Janet Yellen is the Eugene E. and Catherine M. Trefethen Professor of Business Administration in the Haas Economic Analysis and Policy Group and professor in the Department of Economics at the University of California-Berkeley. Dr. Yellen was the Chair of the President's Council of Economic Advisors from 1997–99.

*To our families
and friends*

PREFACE

Blinder's corollary to Murphy's Law: "Economists have the least influence on policy where they know the most and are most agreed; they have the most influence on policy where they know the least and disagree the most."

Economics can make good policy better and prevent bad policy from getting worse—a message frequently controlled by the political realities surrounding federal environmental and natural resource policy. This book is a collection of personal essays by eight former senior staff economists for Environmental and Natural Resource Policy working for the Council of Economic Advisers inside the White House during the Clinton and both Bush administrations. Each essay explores why and how economics matters more to environmental and natural resource policy than many people think. The authors discuss the lessons they learned during the in-house debates over policy inside the administrations during the last fifteen years. The overall goals of the collection are to introduce noneconomists to the economic mindset used in policy debates and to help those economists interested in becoming more apolitical be better advocates of efficiency.

We thank the Bugas, Lowham, and Stroock endowments at the University of Wyoming for supporting the second annual Stroock Forum on

Wyoming Lands and People on "Painting the White House Green," from which this volume originated. We thank Harold Bergman, Keith Cole, Tom Crocker, Mary Budd Flitner, Bruce Forster, Steve Horn, Cynthia Lummis, Terry O'Connor, Jon Roush, Larry Spears, Thomas Stroock, John Turner, the College of Business, the College of Agriculture, and the Department of Economics and Finance for their support. Thanks to the citizens of Centennial for their Wyoming hospitality toward our guests. As always, Jennie Durer was her wonderful self throughout all her organizational and tactical support.

Finally, thanks to the many members of the Council of Economic Advisers whose professionalism over the years has created the environment that made this book possible.

R. L.

J. S.

ABOUT THE CONTRIBUTORS

JOSEPH E. ALDY is a Ph.D. candidate in the Department of Economics and a predoctoral fellow in the Environmental Economics Program at Harvard University. He served at the Council of Economic Advisers in the Executive Office of the President as the staff economist for the environment and natural resources from 1997 to 1998, senior adviser for the global environment from 1998 to 1999, and senior economist for the environment and natural resources from 1999 to 2000.

RANDALL LUTTER is chief economist of the U.S. Food and Drug Administration. As resident scholar, American Enterprise Institute for Public Policy, and research fellow, AEI–Brookings Joint Center for Regulatory Studies, he wrote about a wide variety of health, safety, and environmental policy issues. From 1991 to 1997, he served as economist with the regulatory review office in the Office of Management and Budget, and from 1997 to 1998, he was senior economist for the environment and regulation at the Council of Economic Advisers.

WILLIAM A. PIZER is a fellow at Resources for the Future. His research seeks to quantify how various features of environmental policy and economic context, including uncertainty, individual and regional variation, technological change, irreversibility, spillovers, voluntary participation, and flexibility influence a policy's efficacy. From 2001 to 2002, he served as a senior economist at the Council of Economic Advisers, where he worked on environment and climate change issues.

STEPHEN POLASKY holds the Fesler–Lampert Chair in ecological-environmental economics at the University of Minnesota. His research interests include biodiversity conservation and endangered species policy, common property resources, and environmental regulation. He has served as an associate editor for the *Journal of Environmental Economics and Management*. From 1998 to 1999, he was senior staff economist for environment and natural resources at the Council of Economic Advisers.

JASON F. SHOGREN is the Stroock Distinguished Professor of Natural Resource Conservation and Management and professor of economics at the University of Wyoming. His research focuses on the behavioral underpinnings of private choice and public policy, especially for environmental and natural resources. In 1997, he served as the senior economist for environmental and natural resource policy at the Council of Economic Advisers.

RAY SQUITIERI is senior economist at the Comptroller of the Currency. He was previously a senior economist in the Office of Economic Policy, U.S. Treasury, where his primary responsibility was energy and environmental issues. From 1991 to 1993, he was a senior staff economist at the Council of Economic Advisers, where he worked on environment issues.

MICHAEL A. TOMAN is an adjunct faculty member at the Nitze School of Advanced International Studies, Johns Hopkins University. He also served as a senior fellow and director of the Energy and Natural Resources Division at Resources for the Future until 2003. From 1994 to 1996, he was a senior staff economist at the Council of Economic Advisers, where he had staff responsibility for all the environmental issues facing the council.

JONATHAN B. WIENER is professor of law at Duke Law School, professor of environmental policy at the Nicholas School of the Environment and Earth Sciences, professor of public policy studies at the Sanford Institute of Public Policy, and faculty director of the Duke Center for Environmental Solutions. He is also a university fellow of Resources for the Future. From July 1992 to October 1993, he served as the senior staff economist for environmental and regulatory matters at the Council of Economic Advisers.

LESSONS FROM A HOT SEAT

Randall Lutter and Jason F. Shogren

Presidents, like kings, lead cloistered lives. Surrounding the president is a small collection of loyal friends, advisers, and aides, all dedicated to ensuring that the administration achieves its political goals while enjoying broad and deep political support. Collectively, these people have remarkable control over the information that reaches the president. They write him background memos, schedule his meetings, and decide who gets an audience and who gets the door. From their positions at the right hand of power, they know that most people wanting the attention of the president—but willing to settle for his aides—have an agenda or represent some special interest. All information offered by outsiders is suspect; what matters as much as the message is the identity of the messenger or his or her sponsors. In this charged environment, neutral unbiased information, if timely and on point as well, can be invaluable. The role of the president's Council of Economic Advisers (CEA) is to supply such information to the president.

While the best-known work of CEA involves analyses of the macroeconomy or international trade and finance, the staff of CEA is oriented toward specialists in particular markets, such as labor, health, agriculture, energy, and antitrust and regulation. Environmental and natural resource issues are especially important, at least in terms of occupying the time of key decision makers, because deep disagreements have characterized a broad variety of environmental policies. This was the case within the Clinton administration and still is the case in the broader Washington community. Whatever their source, these disagreements tend to create demand for a timely and independent perspective on policy questions, one that may be seen as authoritative, technically reliable, and free from obvious bias.

Although CEA has taken a severe view of many environmental policy initiatives, to the point where President Clinton even publicly labeled his economic advisers "lemon suckers," neoclassical economists would recognize its positions as proefficiency rather than antienvironment. CEA generally has supported more economic analysis of policy alternatives and greater consideration of tradeoffs in environmental policy debates. The general goal is to help guide public policy to provide more environmental protection at lower cost. And although CEA's recent move out of the Eisenhower Executive Office Building, which adjoins the White House, lessens the impromptu contact, the economists working in the council still try to provide this sound and fair-minded economic advice for the president and senior White House staff. The single overarching thrust is the importance of efficiency—that is, policies that maximize a broad notion of net benefits to society. Of course, people can still offer fair-minded advice based on noneconomic concerns even if it contradicts the opinion of council members.

Implementing this sensible-sounding goal is difficult and vexing for senior policymakers for a variety of reasons. First, environmental policy debates are polarized, with little constituency for cost-effective environmental protection and great demand for symbolic victories. The difficulties in measuring environmental threats and progress give opportunities to demagogues and make it harder to build a consensus for reasonable and practical policies. As students of the history of the environmental movement will immediately recognize, its great successes have been built on public perceptions of imminent environmental crises: silent spring, killer smog, and burning rivers. This pattern provides environmentalists with incentives and, for some, even license to overstate environmental risks. At the same time, regulatory agencies such as the Environmental Protection

Agency (EPA) have traditionally adopted conservative models to assess risks, in part over confusion about the purpose of such models and in part because they help justify the stringent risk-management measures preferred by risk-averse bureaucracies. Industry, on the other hand, has acted boldly in challenging environmental regulations and statutes, in part because EPA, while it has authority over many different industries, has insufficient authority over most to prevent litigation by simple intimidation. As a result of such litigation, the law-enforcement community has tended to prefer regulations that are easy to enforce, as opposed to cost-effective. Indeed, some observers believe that most environmental policy debates are conducted without any single institution consistently advocating efficient solutions. CEA would be a counterexample to such a claim.

The set of essays presented here touches on three broad themes related to the role of economic analysis in environmental policymaking. The relationship between analysis and policy matters, because analysis, after all, is what economists do, and its application to federal policymaking is for some its greatest and most important purpose.

First, good economics matters. Some economists caricature U.S. environmental policy as being generally free from the sort of market solutions that most textbooks suggest are efficient (for example, large-scale tradable permit markets exist only for sulfur dioxide). But without careful analysis to identify and justify efficient options, policies could be even less efficient. The essays by Aldy, Polasky, Toman, and Wiener all argue in favor of the importance of sound and careful analysis in developing good policies. Aldy describes the efforts of CEA in the last years of the Clinton administration to promote the idea that the costs to the United States of complying with Kyoto would fall with greater international permit trading and with participation by developing countries in the international emissions permit markets envisioned by the Kyoto Protocol. Wiener recounts his campaign in the early 1990s to establish a program to promote preservation of forests in developing countries. Without good analysis, government policies can provide for or encourage an efficient use of scarce resources only through guesswork or good luck.

Second, economic analysis, regardless of how it is conducted, is no guarantee of policy decisions that reflect simple ideas of static or even dynamic efficiency. An economist's corollary to Murphy's Law exists: economists have the least influence on policy when they most agree. Lutter, Polasky, Shogren, and Pizer show how a variety of institutional and legal obstacles preclude the adoption of policies consistent with simple ideas of efficiency.

Lutter argues that EPA's 1997 air quality standard for ozone is likely to be infeasible and in any event is too stringent to represent an efficient balance of costs and benefits. Polasky reviews the dominance of politics over economics in influencing policy decisions pertaining to energy restructuring. Shogren also points out that many Washington policy solutions inflict remedies on the West that are worse than the economic problems they purport to address. In general, economists appreciate that White House politicians take into account specific analyses about specific policies and blend them into the overall mix of social policies that touch on but go beyond the environment. Pizer, in an insightful comparison of the Bush administration's very different policies to cap emissions of conventional air pollutants while promoting voluntary measures to reduce global warming, argues that the maturity of institutions and of a specific policy debate is an important and often overlooked determinant of policy outcomes. The authors explore how the role of economic analysis in developing support for efficient policy solutions is limited by institutional and political constraints that vary with the nature and sophistication of public debate about a specific issue.

Third, good economics ultimately affects decision making, as either a positive or a negative force. Aldy and Toman describe economics as a positive force. Economic analysis positively affected the development of the Clinton administration's climate protection policies because it served to identify and highlight the costs of restrictions on international trade in permits and the benefits of developing-country participation in the Kyoto Protocol. Lutter provides a negative example: EPA's incomplete analysis of its 1997 air quality standard for ozone has left it vulnerable to repeated court challenges and leaves fundamental questions about whether it is achievable unanswered. The neglect of economics in developing that regulation has come back to haunt the agency. Pizer contrasts areas where economics was particularly important (design of systems of tradable permits) with those where its role was much more limited (setting emissions caps) and explores at length how a mature political debate is a precondition for sound economic analysis to lead to efficient environmental policies. Squitieri looks at the mutual dependence of analysis and decision making, and shows that there is a precarious balance between economic analysis and policy decisions.

Each essay focuses on an environmental debate in which the author played a key role. In Chapter 1, Pizer contrasts two major environmental debates that preoccupied White House decision makers and newspaper readers during his time at CEA in the early part of the second Bush admin-istration. On one hand, the administration proposed its Clear Skies Initia-tive, a market-based cap-and-trade legislative proposal to lower the emis-sions of sulfur dioxide, nitrogen oxides, and mercury from power plants and other large sources dramatically over the next 15 years. On the other hand, it rejected the Kyoto Protocol to control greenhouse gas emissions, as well as all proposals for mandatory measures, instead endorsing only vol-untary guidelines and an eclectic collection of tax credits for clean energy technologies. Pizer shows how the role of economics is greater in certain aspects of clean air policy, already informed by existing programs and expe-rience with EPA regulations, than in climate change debates, where the novelty of the issue and the relatively ephemeral nature of the damages have forestalled the development of any social consensus about how much or when to control emissions.

In Chapter 2, Lutter sketches how preconceptions about the nature of human effects on the environment prevented sensible development of the EPA's 1997 national ambient air quality standard for ozone. He first shows that EPA's decision-making process is inimical to economic reasoning. The Clean Air Act directs EPA to set standards to protect public health with an adequate margin of safety, irrespective of cost or even feasibility. EPA set air quality standards intending to avoid demonstrated significant adverse effects to health, an approach that appears subjective. Lutter argues that EPA's neglect of costs and feasibility in setting standards appears to have led it to issue ozone air quality standards that are infeasible in several metro-politan areas. Because attaining the standards is each state's responsibility, their infeasibility may lead to chronic noncompliance. Finally, in a telling example of how preconceived notions preclude rational decision making, Lutter shows how EPA's refusal to consider health benefits of ground-level ozone, even when its own staff analyses indicated that these were signifi-cant, left the regulation vulnerable to a court challenge. The D.C. Circuit Court of Appeals ruled in May 1999 that EPA's rule violated the Clean Air Act because EPA had not considered the health benefits of ozone. This shortcoming is still an ongoing policy problem, as EPA's response to the court decision in January 2003 suggests.

In Chapter 3, Toman of the Inter-American Development Bank exam-ines the role of economic analysis in the development of climate change policy. He reviews consensus conclusions about what economically sensi-ble climate policies would look like, including moderate emissions reduc-tion targets that are flexible and increase in stringency over time, flexible compliance strategies, and early and positive engagement with developing countries. The Kyoto Protocol stands in sharp contrast to these conclusions in requiring sharp binding cuts in U.S. emissions over a short period of time. Moreover, it leaves to the future a commitment to international market mechanisms and the issue of developing-country commitments to greenhouse gas emissions limits. Toman next looks at the development of U.S. climate change policy from 1992 through 1998, when CEA chair Janet Yellen testified to Congress about the economic implications of the administration's climate change policies, including its commitment to the Kyoto Protocol. He notes that economic analysis was important in providing a number of useful ideas for increasing the cost-effectiveness of climate control policies, but it also has emphasized the near-term costs of greenhouse gas control, long-term benefits, and the importance of adaptation to climate change as a risk-mitigation measure. He is skeptical that the Bush administration's respect for economic analysis in developing climate policy is an improvement over the Clinton administration.

Whereas Toman looks at the later years of the Clinton administration's climate policy from the outsider's perspective, Aldy in Chapter 4 takes us inside the debate. Aldy reviews the political economy leading up to the Kyoto negotiations in 1997. He stresses the three key factors that mattered inside CEA for providing advice on U.S. climate change policy decisions: economic analysis; flexible, market-based implementation policies; and developing-country participation. The economic analysis conducted inside the administration showed how vital both international emissions trading and developing-country participation were to meet the Kyoto emissions targets at the lowest possible costs. In effect, Aldy stresses that the economic analysis served a vital role in the internal debate by raising the political costs of diplomatic efforts that did not result in either international trading or the participation of big countries outside the protocol, such as China and India.

Wiener next discusses CEA's role in promoting international forest conservation efforts early in the 1990s. He argues that deforestation in developing countries constitutes a negative international environmental externality in that deforestation deprives people in other countries of the value of carbon tied up in trees and of a diverse biological ecosystem necessary for the survival of many species not found elsewhere. Wiener describes the existing institutional arrangements that protect forests in developing countries and also complicate efforts to expand such protection. He then sketches a modest federally financed forest preservation fund, which would pay landowners in developing countries to preserve forests. After modifying the proposal to cap liabilities in accordance with concerns of the Office of Management and Budget, this idea was adopted by the first Bush administration in 1992 and later endorsed by the Clinton administration soon after taking office. This is an example of a new program being so successfully championed by nonpolitical staff in the White House that it survived a change of administration.

In Chapter 6, Polasky describes the policy deliberations over environmental aspects of the energy restructuring debates before the California energy crisis. He shows that economic analysis of the cost-effectiveness of the different proposals served to structure the debate and provide a metric for judging the merit of alternative proposals. Disagreements about the importance of environmental problems within the administration prompted discussions about environmental policy almost to the exclusion of other issues important in energy restructuring. With the benefit of hindsight, this diversion of attention to the environment and away from the importance of proper incentives in the transition to a less regulated market was unfortunate. Problems related to poorly designed incentives are what bankrupted California electric utilities, saddled taxpayers with a fat debt, and nearly caused extensive brownouts throughout the state.

Next, Shogren assesses whether the environmental problems common in the rural interior West all come from the East—that is, Washington, D.C. He suspects that the Clinton–Gore administration pursued centralized environmental policies to gain votes on the coasts, and this suspicion is echoed in the maps of the results of the 2000 presidential elections: Albert Gore did much better on the coasts than in the West—the big red area on the maps on Election Day 2000. Shogren reviews myths popular in Washington about the application of economics to environmental problems. He pops many of them with a bang.

In the final chapter, Squitieri discusses the causes of tension between analysts such as economists and policymakers. After identifying a collection of welfare-improving proposals that Washington has not adopted, he notes that economists, risk assessors, and other analysts acquire knowledge by following a prescribed sort of investigation into the possible consequences of alternative government actions. This scholastic approach, which follows in the traditions of Aristotle and Thomas Aquinas, is incompatible with a mystic approach to the acquisition of knowledge. Mystics, such as Joan of Arc, says Squitieri, acquire information without research according to prescribed rules by relying instead on special insights that may appear to be divinely inspired. In our daily lives, we all make decisions based on both the scholastic approach (is this refrigerator a better buy than that other?) and a mystic approach (do I love this person enough for marriage?). In a somewhat similar way, regulatory agencies must conduct scholastic research into the effects of alternative policy choices, but regulators also need to use special political insights about important intangibles to reach conclusions about which policy is best. Government decision making works best if it involves both scholastic and mystic approaches, and if there is mutual respect between the analysts doing the scholastic work and the policymakers needing to make political judgments based on intuitive insights. It works quite badly, Squitieri says, if people expect the scholastic work of economists to dictate policy decisions, or if regulators subvert scholastic work so that it appears to support preordained political decisions even when more neutral analysis would reach opposite conclusions.

This collection of essays will remain valuable even as the specific players and debates evolve, because environmental protection will continue to be a major political issue. Environmental protection is a central theme in all international summits today, from the environmental summits such as the 1992 Rio meeting and the Kyoto Protocol for climate change to trade negotiations such as the General Agreement on Tariffs and Trade (GATT). Nations will continue to seek ways to reduce habitat destruction in rain forests, mountains of Central Asia, and coral reefs. For years to come, policymakers in Congress and in regulatory agencies will wrestle with how much environmental protection is enough, what it will cost, and what else we could have spent our money on. It is sure to figure prominently in debates in future election years.

The unique perspective that the authors have enjoyed while working for CEA should appeal to readers other than economists. This book will be a valuable part of the libraries of ecologists, biologists, political scientists, and economists working in environmental protection and biodiversity. Economists will find the papers on incentives useful for the way they begin with and depart from the Coase theorem, a cornerstone idea in addressing environmental externalities since 1960. The insight into the real-world political economy contributes to the literature in regulatory economics and political science that seeks explanations for government policies and decision making. At the same time, ecologists and biologists will gain from the knowledge on how decisions are made. The essays put the reader into the middle of the current policy debate on how to proceed with cost-effective environmental policy at a national level.

A new chair at CEA may be surprised at how much attention he or she must pay to environmental policy. Not only are Congress and the White House rethinking air pollution regulation, energy security, climate change policy, and other related issues, but these issues are contentious by their nature.

The White House economists' attention to environmental issues is good news because, as shown in these essays, economists are environmentalists. Although often portrayed in the press as in conflict with environmental activists, economists in government have tended to advocate policies that implicitly assign substantial value to environmental protection, even if that value is sometimes less than what some environmentalists would like. For at least a century, economists have asked society to make choices so that market prices reflect the true social costs of environmental hazards, thereby providing reasonable protection from such hazards. Economists are happy to stand up and say that wealth spent in one place is not spent elsewhere, even in instances where the symbolic value of certain goods—children's health, saving the planet—can easily derail logical debate and deliberations. By helping establish comparisons of costs and benefits as the best measure of the merit of policy, economists have offered their best advice on how to protect and improve human and environmental health, given available resources.

CHAPTER ONE

A TALE OF TWO POLICIES: CLEAR SKIES AND CLIMATE CHANGE

William A. Pizer

On February 14, 2002, President George W. Bush announced two new initiatives: one to reduce emissions of sulfur dioxide, nitrogen oxides, and mercury from power plants—the Clear Skies Initiative—and the other to reduce economywide emissions of greenhouse gases—the Global Climate Change Initiative. Both seek quantified emissions reductions, and both affect fossil fuel use. Otherwise the initiatives are strikingly different. The Clear Skies Initiative aims at absolute reductions through a legislated program of cap and trade, whereas the Climate Change Initiative strives for reductions in emissions intensity (emissions relative to economic activity) through a mostly administrative program of tax incentives, voluntary challenges, and transferable credits for emissions reductions that presumably would be honored under a future policy.

The Clear Skies Initiative was hailed, even by many critics, as a constructive contribution to the debate over new power plant regulation. In sharp contrast, the Climate Change Initiative was assaulted by both the left and the right. Much of this difference in reaction had to do with the differing contexts of the two announcements, not just the substance. Clear Skies represented a better way to approach existing clean air requirements; the Climate Change Initiative pursued an entirely new avenue of environmental protection. An interesting question is whether and how these policies can be viewed as movements toward current or future, efficient or cost-effective solutions—institutional progress in a broad sense. (Efficient policies are those that optimally balance costs and benefits; cost-effective policies are those that, given an environmental goal, achieve the goal at the lowest possible cost.) While economics often leads us to evaluate policies in an institutionally static sense, focusing on immediate costs, benefits, and cost-effectiveness, improving the likelihood or opportunity for efficient and cost-effective solutions in the future can be an equally important goal. Even viewing one or the other policy as an insufficient response in the static sense, arguably both policies achieved this notion of institutional progress.

The vantage point of the Council of Economic Advisers (CEA) offers a useful platform for examining the development and outcomes of the two initiatives. What are the consequences of particular policy choices? Do particular choices represent a step toward or away from cost-effective solutions? Do they make cost-effective solutions more likely in the future? Which concerns other than economics, such as historical experience and politics, are important, and how large or small was the role of economics in the decision-making process? What was really possible?

Consider the broad context of the two initiatives. The Clean Air Act already mandates limits on sulfur dioxide and nitrogen oxides emissions and will require even stricter limits in the near future. Limits on mercury emissions from power plants also will be imposed in the near future by regulations required by the Clean Air Act. The Clear Skies Initiative offers a way to meet and exceed the level of environmental performance likely to be achieved under this current law, but at a lower cost through a more efficient system of tradable permits. Everybody wins relative to the status quo. In contrast, limits on carbon dioxide and other greenhouse gases are entirely new. Such requirements are obviously a burden—in this case a significant burden—on some stakeholders, even if society as whole will benefit. Not everybody wins. (Even though the Climate Change Initiative lacks mandatory emissions requirements, it represents a formal acknowledgment that mitigation is necessary and that quantitative goals are appropriate, raising the expectation of a burden in the future.)

These are not the only contextual differences; important economic, environmental, and institutional distinctions also exist between the regulation of conventional pollutants and of greenhouse gases. Conventional pollutants are typically controlled through a combination of end-of-pipe technologies (scrubbers), process changes (firing design), and some fuel switching (from high- to low-sulfur coal), reducing emissions intensity per unit of energy. Because these approaches are inexpensive relative to overall generation costs, they lead to minor increases in delivered energy costs, and virtually no reductions come from decreases in total energy usage. In contrast, end-of-pipe treatment and process changes are not practical for the control of carbon dioxide, the principal greenhouse gas—unless the carbon dioxide is separated and pumped underground or undersea, a very expensive and, for the near term, impractical endeavor. Therefore, reductions in carbon dioxide emissions intensity can arise only from switching fuels from coal to gas, nuclear, and renewables, and decreased energy use quickly plays an important role. This implies higher—and more visible—costs for significant emissions reductions.

Opportunities for inexpensive reductions in conventional pollutants are concentrated in the power sector, where there is a long history of regulation. For the reduction of greenhouse gas emissions, many inexpensive opportunities exist outside of this sector, requiring the engagement of an entirely different set of stakeholders. These opportunities—which have the same environmental benefits—include energy-related CO_2 reductions in other industries and sectors, reductions in emissions of other greenhouse gases, biological sequestration of carbon dioxide, and the reduction of emissions in other countries.

On the environmental side, the adverse consequences of conventional pollutants are immediate and dissipate relatively quickly once emissions are reduced. In contrast, the consequences of carbon dioxide emissions persist for hundreds of years after emissions occur. Even if all greenhouse gas emissions were cut to zero tomorrow, it would take centuries for the current atmospheric changes (about a 30 percent increase in carbon dioxide) to return to preindustrialization levels. Perhaps more important, the kinds of attention-grabbing events that helped spur the regulation of conventional pollution—a flaming river, a killer smog, or the risk of a silent spring—are unlikely to spur the regulation of greenhouse emissions. At least, not until it is too late to change the outcome.

In the end, these contextual features and economic rationales cannot explain all of the differences between, or elements within, these two policies, a gap filled mostly by political interests. Interest groups tend to weigh in most heavily when numerical targets are being set, when property rights in one form or another are being allocated, and when decisions are made to move closer to mandatory regulations—the last being particularly relevant for the distinction between the Clear Skies and Climate Change policies.[1] In all of these cases, winners and losers are easy to spot, and the stakes are basically zero-sum among environmental and economic stakeholders. The high-level attention given to emissions targets in Clear Skies follows this rationale. On the other hand, the fact that the administration's debate over allowance allocation rules was not so politicized is a little surprising—perhaps it was because this was viewed as a more natural battle for Congress to hash out.[2]

When viewed through the lens of progress toward current and future efficiency or cost-effectiveness, both policies show improvements. The Clear Skies Initiative builds on the successful acid rain architecture, replacing a tangle of existing regulations with a simple approach based on market incentives, including a price-based safety valve, revenue-generating allowance auctions, and reductions in SO_2 and NO_x more in line with expected benefits. Each of these represents movements toward greater efficiency, lower costs, and higher net benefits. The Climate Change Initiative sets a reasonable goal, introduces the notion of an intensity target, and provides a system of transferable credits for emissions reductions. Although it falls short of many people's goal of embracing a mandatory mitigation policy, the announcement includes several interesting and arguably progressive elements. The emphasis on mitigation moves away from a policy focused solely on science and technology. The goal of 4 percent reductions from forecast emissions levels in 10 years is within the range of economically sensible near-term targets. An intensity approach draws attention to the importance of balancing economic and environmental concerns. And the transferable credit system could be the beginning of a tradable permit system and encourage institutions to deal with sequestration and fugitive emissions. Although I am among those who believe a mandatory mitigation policy is warranted and necessary, these steps do represent progress and move the United States closer to an efficient market-based solution.

THE EVOLUTION OF POWER PLANT REGULATION

The 1990 Clean Air Act Amendments established the first significant market for emissions allowances in the world. In 1995, the first year of the program, more than 1,000 transactions valued at more than $10 million occurred. By 2000, this had risen to 10,000 transactions valued at $20 million. Meanwhile, SO_2 emissions have fallen more than 50 percent, exceeding environmental goals over the first five years.

With virtually no precedent at the time, one wonders how the Acid Rain program came about. Key players certainly existed, as did a useful alignment of various political constellations: acknowledged failure of the existing Clean Air Act to achieve desired reductions in SO_2, increasingly strong evidence linking SO_2 emissions with environmental and health consequences, and sufficient technological experience with SO_2 controls to understand with some certainty the likely costs of various emissions targets. In addition, voices were encouraging the use of market forces to achieve environmental goals.[3]

Cohen, Kete, and Stavins discuss the players and forces leading to that remarkable event. After a decade of failures by both the executive and legislative branches of government to deal with acid rain, presidential candidate George H.W. Bush announced he would reduce SO_2 emissions by "millions of tons" in 1988. This announcement culminated years of claims by the Environmental Protection Agency (EPA) under Reagan that acid rain control legislation was unnecessary. It followed decades of debate between northeastern states suffering from the effects of acid rain and midwestern states that burned large amounts of coal.

By the time of the elder Bush's announcement about acid rain, EPA had estimated health and visibility benefits (excluding forest and ecosystem damage) of $3 billion to $40 billion for an 8- to12-million-ton reduction, and costs of $3 billion to $6 billion. Meanwhile, Senators Tim Wirth and John Heinz were popularizing reform ideas. Their "Project 88," written by a team of policy analysts at Harvard University, provided a simple recommendation to the new president. Emissions trading would achieve a chosen environmental goal in a less expensive way.

This alignment of political ripeness, scientific evidence, and popularization of market principles was an unusual congruence, attributable more to time and place than to any simple theory, according to Kete. Less than two years after the 1988 election, the 1990 amendments were passed. And a decade later, presidential candidate George W. Bush found himself revisiting these same issues.

THE FOUR POLLUTANT (4-P) LANDSCAPE IN 2001

By the time of the 2000 presidential election, there was no longer a debate about the use of tradable permits to regulate power plant emissions—both candidates announced support for a system of tradable permits to regulate power plant emissions of four pollutants: carbon dioxide, sulfur dioxide, nitrogen oxides, and mercury. The debate moved on to questions of which facilities would be included, what the emissions levels should be, when they should be achieved, and what restrictions, if any, should be placed on trades. Such a consensus was no small achievement and still stood on rather shaky ground. Fall 2000 also witnessed the energy crisis in California and consequent breakdown of the Los Angeles RECLAIM market for NO_x emissions allowances, giving rise to some skittishness about markets and all their wondrous benefits.[4]

During the presidential campaign, then-governor George W. Bush notably voiced support for regulating power plant emissions of carbon dioxide, as well as conventional pollutants regulated under the Clean Air Act. Candidate Bush opposed the Kyoto Protocol as an expensive burden on the United States that would do little to curb global emissions, as large developing countries were exempt from any obligations. Carbon dioxide emissions limits for power plants, however, could be much more modest in nature while showing a tangible effort to address climate change. Meanwhile, as the electricity sector invested billions of dollars in new generation capacity in the coming years, it would give them greater certainty about the future regulatory context—one that was increasingly likely to regulate greenhouse gas emissions.

Immediately after assuming office, President Bush's opposition to the Kyoto process appeared to soften, with EPA administrator Christie Whitman stating on March 3, 2001, that the United States would seek to make the Kyoto Protocol "workable."[5] That changed abruptly on March 13, when the president announced that he continued to oppose the Kyoto Protocol and, further, did not believe that carbon dioxide should be regulated in new multipollutant regulation. The president's announcement, in response to a letter from Senators Chuck Hagel, Jesse Helms, Larry Craig, and Pat

Roberts, sparked a tremendous reaction domestically and especially at the international level, mostly negative. This set the stage for two major environmental policy developments: a three-pollutant initiative for regulating sulfur dioxide, nitrogen oxides, and mercury—but not carbon dioxide—and a new climate change policy that moved away from the Kyoto Protocol.

In retrospect, the reaction to the March announcement was as much about how the announcement was made as it was about the substance. Many of our partners in the umbrella group that worked together on the Kyoto Protocol were caught entirely by surprise by the clear dismissal of the protocol. Similarly, the domestic audience was caught off-guard by the change vis-à-vis the campaign promise on power plant regulation. Although the policy decisions themselves were perhaps inevitable, a more nuanced announcement could have smoothed over some of the negative reactions.

PURSUING CLEAR SKIES

When work began in earnest on the Clear Skies Initiative in spring and summer 2001, a tremendous number of policy questions remained unanswered regarding emissions sources covered by the program, emissions schedules, trading rules, allocation, and interaction with existing programs and Clean Air Act requirements.[6] The development of sensible answers to these questions required economic analysis along with extensive discussions among the EPA, the Department of Energy (DOE), the Department of the Interior, and multiple offices within the Executive Office of the President—the Office of Management and Budget (OMB), the National Economic Council, the Vice President's Office, the Chief of Staff's Office, the Council of Economic Advisers, and the Council on Environmental Quality (CEQ). Staff frequently met a couple times each week, and meetings at the deputy level were almost weekly from the summer through the following spring. Principals, such as cabinet members, met when necessary to discuss and resolve the thorniest issues, with debate on specific emissions schedules continuing virtually until the policy announcement on February 14, 2002.

Several forms of economic analysis were involved. Very detailed analysis of policy consequences for electricity generation, prices, fuel use, emissions, and pollution concentrations figured prominently in the selection of emissions reduction schedules. This analysis was based on modeling at EPA and

DOE. Experience with other programs and results from the economics literature were brought to bear on the design of the trading system. For example, how should the policy deal with uncertainty about costs? What would be the economic consequences of alternative allowance allocation schemes? If auctioned, how should allowance auctions be structured? How should a subsidy to encourage early technology adoption be designed to achieve the most adoption at the least cost? These kinds of analyses were more qualitative. Finally, it was important to consider the interaction with existing regulatory programs. Clear Skies was supposed to replace a cumbersome web of regulations at the federal, state, and local levels with a simple market-based approach. How much of this replacement needed to be explicit and how much could be implicit because the proposed limits made it unnecessary for state and local governments to seek further reductions at participating facilities? This analysis was both quantitative and qualitative. The role of CEA was both to produce analyses and to help arbitrate when different analyses—often arising at EPA and DOE—came to different conclusions.

These analyses often guided decisions. Consistent with earlier observation that political interests tend to weigh in when decisions are zero-sum, economic analysis faced the greatest scrutiny and competition from other concerns in the arena of setting numerical emissions limits and in allocating emissions allowances.

Emissions Sources

The identification of sources covered by the proposed Clear Skies legislation was both one of the first and one of the last topics debated by the interagency group. Economic reasoning suggested that a trading program should include all emissions sources with similar emissions characteristics and control opportunities. That is, one might exclude small, difficult-to-monitor sources, as well as those that faced high control costs. But otherwise, broad inclusion ensured that all inexpensive reduction opportunities were encouraged, achieving a given level of emissions reductions at the lowest cost. In the case of Clear Skies, sources would seem to include not only power plants, but also industrial boilers, which have many of the same control options (and costs) as electric generating units but face less regulation, presumably giving them lower marginal costs.[7]

It might seem counterintuitive to argue that only those sources with similar characteristics should be included. The whole idea of trading is to bring *all* sources together and let those with high- and low-cost abatement options trade with each other. Indeed, if all sources have similar costs, is trading even necessary? This ignores a few practicalities, the most obvious being monitoring and observation of emissions. It may not be practical to include some sources, especially small ones, in a program that requires expensive monitoring equipment. Less obvious is that it may be unnecessary for a cost-effective solution to include high-cost sources if the effect on output prices would be small.[8] High-cost sources will never abate—they will always buy allowances—and their inclusion in a regulatory scheme potentially adds another opposing voice to the debate with little effect on emissions. This suggests that the key element for a least-cost solution is to include the practical, low abatement cost sources in the program, but not necessarily all sources. Note that even if all sources have similar characteristics, trading occurs because the allocation may not reflect actual need and because facilities are unlikely to reduce emissions simultaneously; some may do so first and sell excess allowances to others.[9]

In addition to economic sense, inclusion of industrial boilers also seems to make some political sense. Once power plant reductions have occurred, industrial boilers will be the largest source of SO_2 emissions. As such, they likely will face increasing scrutiny by those states that remain out of attainment even with dramatic emissions reductions from power plants. By giving them a relatively generous allowance allocation, they could be better off and opportunities for cheaper reductions would be brought into the program. This argument was based on the belief that many of the reduction options followed by power plants under the Acid Rain program could be adopted by industrial boilers at about half the cost of the deeper cuts being proposed under Clear Skies. That is, while the Clear Skies emissions limits will require scrubbers on every power plant, industrial boilers could reduce emissions relatively cheaply by switching to low-sulfur western coal, the compliance route taken by many power plants under the current program.

Despite these arguments, it became clear early on that the stakeholders associated with industrial boilers did not see it this way. Their perspective was that further regulation under the existing Clean Air Act was not as imminent as members of the interagency group might have believed. Further, they believed that their future under the existing system—striking their deal separately from the power plants or hoping reductions would be delayed beyond the Clear Skies schedule—was more promising than inclusion in Clear Skies. Because it did not make sense to begin the public debate with one more adversary at the table, and with only economists likely to chime in about the efficiency gains, industrial boilers were not included in the universe of affected sources.

The issue of covered sources arose once again when attention turned to the potential voluntary opt-in of industrial boilers.[10] This discussion took place after the president's February 14 announcement, but before the administration released details of the initiative. Under a voluntary opt-in, facilities that are not required to participate in the trading program can voluntarily decide to do so based on established rules. Here, the debate surrounded what kind of deal might be offered to entice industrial boilers to voluntarily participate. One approach would have been to offer a generous package, encouraging more facilities to participate and bringing down costs. A more subtle concern was that many of those choosing to participate might still have reduced their emissions absent the voluntary program (often referred to as "anyway tons" of reductions, because they would have happened anyway). With the opt-in program, these reductions would generate allowances for other sources to increase their emissions.[11] A more generous allocation to sources voluntarily opting into the program would lead to even greater emissions increases at existing sources. Under this line of reasoning, a more modest opt-in package would have been appropriate, forgoing some low-cost opportunities at industrial boilers in favor of safeguarding the emissions limit.[12]

In the end, the proposed legislation followed the second line of reasoning, with a relatively modest allowance allocation equal to half the source's historic emissions level. The rationale was that relief from other, onerous, and inefficient regulations would be the primary mechanism for encouraging voluntary opt-ins.[13] This allocation remained consistent with the original levels contemplated when industrial boilers were going to be fully included in the program.

Emissions Reduction Schedules

Easily the most controversial and contentious question surrounding the Clear Skies Initiative concerned the level and schedule of required reductions. The process was confused by the elusive question of criteria for judging an appropriate schedule. Beyond the question of criteria, at various points there was concern that the proposed target might not be the eventual target legislated by Congress, suggesting a more strategic approach, but this concern was downplayed. Economics offers a straightforward criterion: balance costs and benefits when choosing an emissions target. In particular, continue to reduce emissions until the marginal cost of further reductions equals or exceeds the marginal benefits.

Unfortunately, that is not the way environmental law is written in the United States. For both ozone (of which NO_x is a precursor) and particulates (to which NO_x and SO_2 contribute in varying degrees), the EPA administrator is required to set ambient air quality standards that, with an adequate margin of safety, are requisite to protect the public health. There is no mention of costs, only benefits. With that in mind, the benchmark for judging the appropriate NO_x and SO_2 emissions schedules for Clear Skies became a mixture of two elements. First, how many areas were being brought into attainment with the ambient air quality standards? This directly responded to the goals of the Clean Air Act and represented an absolute metric. Second, how did the emissions outcomes compare with what would have happened anyway—a "Clean Air Act as usual" scenario? This provided a relative comparison of Clear Skies versus no Clear Skies and focused on the idea that emissions trading was a better way of doing business—better outcomes at lower cost.

It was a slightly different story with mercury, for which the guiding force was the requirement of maximum achievable control technology (MACT) for power plants, based on mercury's designation as a hazardous air pollutant (HAP). In this case, after HAP determination, neither costs nor benefits matter, and EPA is required to set regulations based on what is achievable, achievable being defined by the average control rate of the best-performing 12 percent of existing sources. By requiring power plants to reduce mercury emissions under Clear Skies in a way that was consistent with estimates of likely MACT requirements, the idea was to remove MACT requirements and simplify power plant regulation.[14]

Presumably, electric generators get something out of a multipollutant approach: certainty about emissions requirements and flexibility through markets. They are specifically excluded from most of New Source Review and the mercury MACT. It should be possible, therefore, to maintain their support even while reducing emissions below what would otherwise occur. There is a limit, however. And though the Clear Skies Initiative was clearly an effort to make the Clean Air Act requirements more efficient, it was not an effort to significantly change the burden on power plants.

The problem with these benchmarks for evaluating potential emissions limits—even marginal costs against marginal benefits—is that they all involve many subjective elements. Forecasting likely emissions levels absent Clear Skies is perhaps the most contentious. This contention arises because it is unclear how quickly the Clean Air Act would reduce emissions. At issue are how fast is deemed "practicable";[15] how quickly the agency and other parties will respond to nonattainment caused by upwind emissions sources in different jurisdictions;[16] how aggressively firms, environmental groups, and other stakeholders will challenge rulemakings in court; how the courts would respond; and how power plants in particular would be affected by these decisions.

Disagreement over these baselines was the foundation for the debate over the emissions limits. EPA took the position that rapid reductions were likely to occur under the Clean Air Act, and that tighter caps were warranted given that baseline. DOE took the position that scarce capital and installation capacity would make rapid reductions impossible, and that costs were higher than EPA indicated, thereby recommending looser caps and longer time frames. Some of the disagreement could be traced to variation in modeling assumptions regarding growth and costs, but much of it reflected deeply different beliefs about the future. While providing a framework for evaluating the cost and impacts of competing proposals, there was little role for economics—and indeed, most of the other agencies—to resolve questions of how the regulation under the Clean Air Act will unfold. In the end, the interagency process resolved the disagreement, with limits in the year 2018 of 3 million tons of SO_2, 1.7 million tons of NO_x, and 15 tons of mercury.[17]

Ironically, it was the difficulty associated with the creation of a "Clean Air Act as usual" baseline that eventually allowed opinions about costs and benefits to work their way into the discussion. Within the bands of ambiguity about baseline emissions, the tendency was to be more aggressive regarding sulfur dioxide and nitrogen oxides compared with mercury, based on more convincing evidence of the human health benefits, the linkages to power plant emissions, and the costs of the emissions reductions. It was also confusion about this baseline, created in part by a poorly labeled presentation slide, that led to accusations that Clear Skies would increase emissions relative to forecast levels under current Clean Air Act regulations.[18]

Even as EPA analysis shows that benefits greatly exceed costs under the chosen Clear Skies targets, that same analysis can be used to examine whether the limits might be improved at the margin. That analysis estimates benefits associated with SO_2 and NO_x reductions of $16 billion to $100 billion in 2020.[19] It also estimates reductions in SO_2 of 4.6 million tons and NO_x of 2.3 million tons.[20] A crude division of total benefits by total reductions yields a low value of $1,900 per ton. Prior analysis by

OMB suggests benefits of $500 to $2,300 for NO_x and $3,700 to $11,000 for SO_2.[21] Given estimated marginal costs in 2020 of $1,200 per ton for SO_2 ($700 in 2010),[22] this suggests that a further tightening and more rapid implementation of the SO_2 caps would increase net benefits. The NO_x caps, with a $1,400 to $1,800 marginal cost (in the eastern region) over the period 2010–2020, appear roughly optimal.[23]

Meanwhile, no benefit estimates exist for mercury reductions.[24] Nonetheless, power companies will be required to spend from $1,000 to $2,187.50 per ounce to reduce mercury emissions.[25] Whereas with conventional air pollution the only practical way to reduce human exposure is to reduce emissions, risks from mercury can be reduced by encouraging populations at risk to avoid eating contaminated fish.[26] This leaves a pure cost–benefit analysis at a loss to recommend a particular target. It also draws attention to the fact that significant resources are being spent to obtain an elusive benefit. Should resources be focused where the benefits are known?

The targets and timetables were the most heavily debated elements of the package. Cost–benefit analysis played a role—the overall benefits do exceed the costs by any measure—but its role in narrowing down the choice of targets was at the margin. More detailed economic analysis of the distribution of costs and impacts concerned everyone, as did input from various stakeholders and agencies. But the dominant concern in deciding among competing proposals was meeting or exceeding expected air quality under current Clean Air Act implementation. No doubt these targets and timetables will continue to be debated along similar lines as Congress takes up the proposal.[27]

Trading Rules

Debates over emissions limits offered less opportunity for real economic input, because they were driven in large part by alternate opinions about business as usual under the Clean Air Act, voiced most strongly by EPA and DOE. In contrast, there was no similar conflict over trading rules, and here economic analysis played a significant role. In particular, several lessons that were learned since the implementation of the Acid Rain program were put into practice in Clear Skies concerning price spikes and banking.

During winter 1999–2000, the market for NO_x RECLAIM permits in the Los Angeles Air Quality Control Region completely broke down. Facing a well-publicized electricity crisis, unexpected use of gas-fired generators created a demand spike that pushed prices from $2,000 per ton to more than

$90,000 per ton. Despite the health benefits of reduced emissions, it seems ridiculous to suggest that reductions costing $90,000 per ton are desirable. Scrutiny of the RECLAIM program revealed several concerns.[28]

First, the program did not allow significant banking across years; firms could not save unused allowances for future use. This prevented businesses or speculators from building up a reserve that could accommodate unexpected shocks. It also made the end of each compliance period—when supply and demand had to roughly match—a particularly volatile time. Second, the program did not provide any form of automatic stabilization when prices became astronomical. For example, trading on the New York Stock Exchange is automatically halted when stock prices fall a certain amount. Similarly, the Federal Reserve stands ready to intervene in various bond and currency markets to offset any sudden shifts in demand.

The Acid Rain program successfully dealt with the first concern by allowing full banking of allowances from year to year. This not only provides flexibility, but also encourages efficient reductions over time. Early, inexpensive reductions were encouraged because they created a surplus of allowances that could be banked for the future, when reductions were likely to be more expensive. The Clear Skies program not only allows full banking of allowances from year to year, but also allows the banking of pre-Clear Skies allowances into the Clear Skies program. That is, extra reductions in 2009, the last year under the Acid Rain trading program, result in excess Acid Rain allowances that can be banked into Clear Skies allowances in 2010. In this way, reductions are achieved earlier and at the lowest possible cost.

The second concern, unexpectedly high prices, has not arisen in the Acid Rain program, in part because firms continue to hold a substantial volume of banked allowances. This bank, equal to an entire year's worth of allowances, allows the market to absorb significant fluctuations in annual demand without dramatic changes in price. Nonetheless, it remains a concern as the bank gradually declines, given the RECLAIM example. For that reason, Clear Skies uses a safety valve mechanism to ensure that allowance prices are never driven to astronomic levels. The safety valve establishes a price ($4,000 per ton for NO_x and SO_2 and $2,187.50 per ounce for mercury) at which an unlimited number of additional allowances will be sold. These allowances are subtracted from future allocations—so the environment will be made whole in the long run.[29] In the short run, this provides a guard against extreme price spikes and inefficient use of resources to meet a dramatic short-term shortage.

An interesting debate arose in the interagency process about the appropriate level for the safety valve. The economics literature suggests that if the marginal benefits are relatively flat, a price mechanism is preferable on welfare grounds—with the price (or safety valve) set at the marginal benefit level.[30] Here, there is a bit of a conundrum. The basis for the trading program is a cap designed to meet or help in meeting an air quality standard where costs and benefits no longer play a role (despite evidence that benefits, however uncertain, are relatively flat). At the same time, common sense dictates that there is an unacceptable cost threshold beyond which people agree that costs are too high for those particular reductions.[31] Based on the notion that the current air quality standards and economic analyses of control costs represent the agreed basis for the program, the working group decided to set a safety valve that became relevant only if the allowance prices exceeded normal variability about the expected price—roughly two to four times the expected price. In this way, the market would be expected to operate normally unless the analyses were really off the mark.[32]

Allocation

Other than the schedule of emissions reductions, no question attracted more discussion than that of how to allocate emissions allowances. Several broad approaches were considered: distributing allowances freely to electricity generators using a fixed rule based on historic information, distributing allowances freely to electricity generators using a dynamic formula based on continually updated production information, and auctioning allowances. The economics literature over the past decade has made two relevant points regarding allocation. First, updating allocations tends to distort prices and can substantially raise the cost of a policy.[33] Second, auctioning allowances and using the revenue to cut other taxes can substantially reduce the cost of a policy.[34]

In addition to these general observations, several specific observations are relevant for regulating SO_2, NO_x, and mercury. SO_2 and mercury are emitted by coal-fired plants and not gas-fired plants. Because gas-fired plants frequently provide peaking capacity, coal plants will not be able to pass on those control costs through higher electricity prices in markets where generation is competitive and gas-fired capacity is the marginal cost producer. In those markets, costs must be borne by the owners of the coal-fired facilities, and possibly employees and other input suppliers. In addition, little additional coal-fired capacity is forecast, and this group of stake-holders is relatively static.

NO_x, on the other hand, is emitted by both gas- and coal-fired plants. Therefore, whether coal or gas facilities set the price of electricity in a given market, the cost of NO_x controls and allowances can affect electricity prices, and a smaller share of costs is borne by facility owners, workers, and input suppliers. Also, this group of stakeholders is changing, as new gas facilities are forecast and those facilities will be required to participate in the NO_x market.

The policy discussions also recognized that design choices for Clear Skies likely would have a strong influence on the design of future programs for some time to come—just as the Acid Rain program has influenced program design since 1995. Be they other regional air pollution programs, water programs, greenhouse gas programs, or something else, the legacy of current choices is important for both arresting bad ideas and promoting good ones. This was a real concern at CEA, and our chapter that year in the *Economic Report of the President* focused on institution building. Well-designed policy is an important institution.

From the outset, the economically focused agencies—OMB, CEA, and the Treasury Department—consistently encouraged auctions, based on the efficiency arguments raised in the economics literature coupled with the precedent for future policies. Meanwhile, DOE, the Interior Department, and EPA all emphasized that real control costs already were being imposed on firms, and that the additional burden of purchasing allowances would be unfair. All agreed that this was primarily a short-term argument; over time, capital depreciates and adjusts. With competitive markets, costs are eventually borne by electricity consumers, and owners receive a market return on investment. From this viewpoint flowed the idea of an allocation system that began by giving allowances away freely and gradually evolved toward an auction.

Even more than pushing auctions, OMB and CEA discouraged dynamic allocation based on future production, referred to as an "updating" allocation. In addition to the potential economic inefficiencies, updating creates considerable political divisiveness among electricity generators regarding who would be entitled to the production-based allocations. Economics suggests that under an updating system, the allocation should be broad, including nonfossil as well as fossil generation, in order to minimize what is effectively a subsidy on generation (i.e., giving out valuable allowances in proportion to generation). Focusing this subsidy solely on fossil generation would create a distortion between fuels favoring fossil energy.[35] Even without these economic arguments, nuclear and renewable generators would not be enthusiastic about an effective subsidy for fossil generation. Meanwhile, fossil generators were unlikely to support a system that gave allowances to facilities that were not even required to hold allowances. In the end, these concerns about the divisiveness of an updating allocation were as much a factor in its rejection as the economic arguments.

The evolution toward an auction also reduced some of the arguments in favor of an updating allocation, especially regarding NO_x allowances. Because of the new entry of gas-fired plants and their need for NO_x allowances, there was a sentiment that these allowances would need to be allocated via a formula that was updated to include new entrants and generation, rather than given out on the basis of historic performance. This sentiment was grounded in the notion that it was unfair for some generators to be given allowances freely and for others to have to buy them, giving grandfathered facilities a production cost advantage. Such notions were debunked by arguments that competitive markets would capitalize the value of any permanent allowance allocation into the value of the company owning the allowances, and that there would not be a production cost advantage associated with owning, rather than purchasing, allowances. This led to a second, more subtle concern: the capitalized value of any permanent allowances would reduce the cost of raising capital and still be an advantage that incumbents would enjoy over new entrants. In either case, auctions would remove the perceived inequity by treating all facilities the same way.

Practical arguments against a fixed allocation also helped push the consensus toward an eventual auction. Once a permanent allocation exists, it creates a considerable amount of inertia to maintain the same allocation pattern even as policies are revised in the future. Existing allowance holders have a valid interest in receiving new allowances if the old ones are eliminated. Meanwhile, as circumstances change over time, revised regulations likely will require addressing the needs of a new group of stakeholders affected by the changing regulations. The combination of allowance holders as well as affected sources vying for the same pool of allowances will inevitably make policies harder to adjust. This is related to an observation surrounding the spectrum auctions conducted by the Federal Communications Commission (FCC). Even if one wanted to give away the spectrum, it is hard to identify reasonable criteria on which to base such a giveaway. In the case of pollution, historic patterns provide a basis in the near term, but in the long term, such a fixed basis becomes irrelevant. The auction avoids what would otherwise be an arguably arbitrary allocation in the future based on an antiquated formula.

Early discussions considered separate kinds of allocation programs for each pollutant, including an updating allocation for NO_x, but a consensus gradually emerged. Initially allowances would be given out to existing facilities (as well as soon-to-be-existing facilities) based on historic data. This grandfathering would be done in a similar manner to the Acid Rain program. For SO_2, the allocation would be based on the existing Acid Rain allocations, with some adjustments for newer facilities. For NO_x and mercury, allocation rules would apply particular emissions factors to historic production levels in order to define an allocation share. In both cases, the initial grandfathering would gradually transition to an auction.

One of the few sticking points was the length of transition from grandfathering to auctions. The Interior Department and most other agencies believed that too rapid a transition would cause the auction to be dropped completely during negotiations. OMB focused on the earlier economic arguments that costs are reduced by an auction where the revenues are used to cut other taxes—and wanted to get there quicker. CEA took the position that maximizing the chances that the auction would survive future negotiations was more important than getting there quickly. Initial debate ranged from 40 to 60 years; the resulting consensus was 52 years.

Despite the consensus and eventual inclusion of an auction provision in the Clear Skies legislation introduced in Congress, the fate of the auction provision is unclear today. Considerable opposition continues to exist among power plants and related stakeholders,[36] and few other than economists have voiced strong support for auctions. Meanwhile, there seems to be little incentive for the administration or Congress to fight for auctions: the revenue is small by government standards, and the political benefit even smaller. At least as a starting point, however, the economically efficient solution is on the table.

Interaction with Existing Regulations

Much of the motivation for pursuing the Clear Skies Initiative—as well as the other multipollutant bills, including Jeffords–Lieberman—was to replace the existing tangle of environmental regulations that affect power plants with a simpler, more flexible, and more effective approach. Thus, it was essential to consider the interaction with existing regulations. Here, the major concerns were New Source Review, the process of State Implementation Plans, and the potential for Section 126 petitions.

New Source Review (NSR) is a contentious program under which facilities undergoing major modifications, as opposed to routine maintenance, are subject to strict emissions controls if the modification increases emissions. The gray area surrounding the definition of a major modification, as well as disagreement over the interpretation of emissions increases, has led to considerable litigation and arguably the avoidance of efficiency-improving modifications at some facilities. By avoiding such modifications, facilities reduce the risk of triggering a review and the potential for stringent emissions control requirements.[37]

Few would argue against replacing NSR with a tradable permit program such as Clear Skies. With guaranteed environmental outcomes and none of the perverse incentives created by NSR, such a program is better for everyone. It is cheaper and more effective. The only real question is whether one believes the emissions limits in a program such as Clear Skies are adequate based on what would likely occur under NSR and other programs—a debate over the targets. Based on this reasoning, it was a straightforward decision to exempt electricity generators covered under Clear Skies from most of NSR.[38]

A more interesting question surrounds the State Implementation Plans (SIPs) and Section 126 petitions. Traditionally, states have had the primary responsibility for meeting air quality standards, with SIPs being the primary tool through which the states required sources to reduce emissions in their jurisdictions. A Section 126 petition is a vehicle through which downwind states ask EPA to require reductions in upwind states. EPA articulated a concern that the federal government not preempt states' rights to regulate sources in their own states. At the same time, DOE expressed the sentiment that it was unfair to power plants for states to "take a second bite of the apple" by hitting them with additional requirements after they complied with Clear Skies.

Simple economics, as well as common sense, suggests that for local pollution problems, local responses are necessary and a federal approach will be inefficient.[39] But SO_2 and NO_x tend to cause regional problems because of the high stacks on power plants and chemical reactions involved in creating ozone and fine particulates, suggesting something between a federal and a local solution. Furthermore, modeling suggests that a national program actually reduces emissions where air quality problems will exist—

basically because those are the sources currently having inadequate controls and facing the lowest abatement costs. Only for NO_x emissions, where power plant emissions in the West were typically not significant contributors to air quality problems, did it make sense to split the trading system into regions, with tighter controls in the East.

The general concern about balancing local versus federal authority was resolved by recognition that with Clear Skies in place, the incentives for states to go after power plants a second time would be quite low. First, many more areas would be in compliance with the Clean Air Act requirements than without Clear Skies. Those areas that are forecast to be in attainment based on EPA modeling of Clear Skies are presumed by the federal government to be in attainment through 2015 (technically, they are defined as "transitional"; see Section 3 of Clear Skies). Second, the marginal cost of control at all power plants would be such that little opportunity would exist for further efforts—even if states wanted to impose further controls. With this recognition, no explicit exemption from SIPs was included in Clear Skies. The potential for states to file Section 126 petitions was also retained, but action against power plants was delayed until 2009, when Clear Skies comes into effect, and compliance and implementation deadlines for such petitions were extended beyond 2011.

A different set of regulations known as MACT governs mercury. MACT requires that emissions sources in a given category control hazardous air pollutants such as mercury at a level equivalent to the average rate of the best 12 percent of sources in that category. Because source categories can be defined narrowly—for example, based on the type of coal or firing design—it is unclear exactly what this would mean for power plants. Further, MACT does not define a target concentration or emissions level for hazardous air pollutants—it simply stipulates "maximum achievable" reductions.

Clear Skies interprets the outcome under the mercury MACT and replaces it with a tradable permit system. As was the case with SO_2 and NO_x, the controversy was not so much over the replacement of the cumbersome regulations with a cap-and-trade program as over the chosen level of the caps. Controls on mercury have attracted some attention because under alternative proposals, such as the bill put forward by Senators Jim Jeffords and Joe Lieberman, the allowances are not tradable. Such a stipulation is based on the assumption that mercury deposition is local and therefore trading risks creating mercury hot spots; it is also closer to the MACT requirement of achieving a particular standard at every source. On the other hand, such an approach ignores the heterogeneity of emissions sources and the uncertainty surrounding the public health effects of local deposition.[40]

Technology Incentive Program

As was the case with the introduction of the original Acid Rain trading program, Clear Skies provides an incentive for technology solutions to SO_2 emissions reductions versus fuel switching. The political rationale was the need to support the interests of high-sulfur coal producers in the East. The economic rationale was that it would be inefficient to encourage power plants to switch increasingly to low-sulfur western coal as a cheap, near-term compliance option before eventually installing scrubbers (or other technology options). With scrubbers installed, cheaper, high-sulfur eastern coal would be attractive once again. The inefficiency centered on the possibility that if closed, eastern mines would not reopen.

The economic agencies, CEA and OMB, were generally unmoved by the economic arguments. If power plants knew that the mines would close as they shifted to western coal, and that they eventually would want to return to eastern coal, they could choose to support the mines in order to keep them open. If they did not choose to, this suggests that closing the mines would not truly be inefficient.

Seeing the technology incentive program as a relatively small economic inefficiency but relatively large political concern, CEA and OMB shifted from opposing it to designing the incentive to encourage as much new technology as possible and to cap the cost of the program. Early proposals suggested that a fixed number of allowances be provided per ton of scrubbed SO_2 capacity, without an explicit limit on the total subsidy and an unclear incentive to actually scrub (versus building scrubber capacity without actually using it). The final program set aside a fixed number of allowances and established a reverse auction to distribute them to power plants wanting the subsidy. Under the reverse auction, modeled after a similar program for renewable energy in California, plants would submit bids of annual tons of SO_2 to be scrubbed with new scrubber capacity along with the number of allowances they want to be paid in order to do so.[41] The lowest requested "prices," in requested allowances per scrubbed ton, would be accepted. Every accepted bid would get the highest accepted price, and the winning price would be chosen to exactly use up the 250,000 allowances set aside for the program. This approach would successfully limit the size of the program while guaranteeing that it would achieve the maximum amount of scrubbed SO_2 possible.

Economics and Clear Skies

Around the time Clear Skies was being developed, Howard Gruenspecht (at RFF) and Robert Stavins (at Harvard) wrote an op-ed in the *Boston Globe* arguing that command-and-control regulation in the NSR–SIP style should be replaced by tradable permit programs in the Clear Skies vein. Although the debate over market-based approaches is far from over, it is remarkable that a decade after the 1990 amendments and the introduction of the Acid Rain trading program, the major obstacle to further reform is disagreement over targets and the inclusion of carbon dioxide (the principal difference between the Jeffords–Lieberman proposal and Clear Skies). The legacy of events over the past two decades is that command-and-control regulation is almost uniformly frowned upon, and every policy needs to "harness the market" in word if not in deed. At the same time, environmental law is founded on fundamentally uneconomic concepts such as air standards set to "protect the public health" or "maximum achievable control technology."

The Clear Skies result, arising in the context of a political and analytical process, is a legislative proposal that moves toward increasingly more efficient and cost-effective policy, but leaves room for improvement. Important progress includes the replacement of NSR and the mercury MACT with market-based programs, the increased use of auctions to allocate allowances, the adoption of a safety valve to guard against allowance price spikes, and marginal costs more in line with marginal benefits. Areas for further improvement include the cap levels and novel concepts like interpollutant and intersector trading. Admittedly, some of these reforms seem complicated, but they could offer additional efficiency gains if they gained the support of the constituency groups (beyond economists) necessary to get them enacted.

CLIMATE CHANGE POLICY

After carbon dioxide was peeled off from the multipollutant legislation, carbon dioxide and climate change policy went a different way from Clear Skies. This began around the time I arrived at CEA. On June 11, 2001, the president delivered a speech in the Rose Garden reasserting the United

States' commitment to addressing the threat of climate change, specifically to the UN Framework Convention on Climate Change (UNFCCC), and more generally to further scientific and technological research. There was a sense that this was a first speech—framing the problem and describing a commitment—and that another speech would follow. Indeed, as the State Department team departed for the continuation of the 6th Conference of the Parties (COP) to the UNFCCC in Bonn one month later, there was an unspoken expectation that a future policy announcement was forthcoming. This announcement, at first expected around the time of the next COP in November, was delayed by the events of September 11 and eventually was made in February 2002.

To understand why limits on carbon dioxide went a different way, one must consider the context of climate change policy vis-à-vis Clear Skies. Clear Skies addressed conventional pollution already regulated under the Clean Air Act. Carbon dioxide is currently unregulated (although several states recently sued EPA, asking that it set national ambient air quality standards for carbon dioxide, claiming that CO_2 is a pollutant under the Clean Air Act).[42] Emissions of NO_x, SO_2, and mercury can be reduced through end-of-pipe treatment—scrubbers, catalysts, or filters. Ignoring untested options to sequester carbon dioxide in geological formations or deep underwater, carbon dioxide emissions can be reduced only by switching fuels or reducing overall energy use. The scientific evidence on NO_x and SO_2 allows the construction of broadly accepted estimates of human health consequences and, with assumptions about the valuation of morbidity and mortality, estimates of the monetary benefits associated with emissions reductions. Scientific evidence indicates that human emissions of CO_2 are likely contributing to changes in the climate, but the timing, magnitude, and valuation of those consequences are largely, if not fundamentally, uncertain.[43]

In addition to these differences, it is also important to understand how energy policy choices influence climate change policy choices. In May 2001, the administration released the *National Energy Policy*, which contained an entire chapter on the environment, another on improving efficiency, and another on renewable energy. At the same time, a basic tenet is that reliable and affordable energy is an important ingredient for economic growth and prosperity. When that is interpreted to mean that rising fossil energy prices are undesirable, such a tenet is at odds with the notion of limiting carbon dioxide emissions through a tradable permit system. Unlike a tradable emissions program for SO_2, NO_x, and mercury at power plants, where reductions would mean little consequence for consumer energy prices, a tradable permit system for CO_2 would succeed only by raising fossil fuel prices based on their carbon emissions, thereby encouraging fuel switching and conservation. Such a policy could be designed to encourage modest reductions through a small price increase—but it is still a price increase in fossil-based energy.

The combination of scientific uncertainty, the absence of existing requirements, and the fact that reductions in carbon dioxide emissions require absolute reductions in energy use or fuel switching all led to a distinctly different approach to climate change. In particular, the climate policy focused on more modest reductions than those called for by Clear Skies, as well as an emissions target indexed to economic activity—in other words, emissions intensity. The climate policy also avoided mandatory controls in favor of tax incentives and voluntary programs.

The Process

Whereas the Clear Skies process was guided by deputies with work done at the staff level at various agencies, the Climate Change Initiative was guided by principals with work done primarily at CEA and CEQ. This reflected the difference between the broad, high-level policy debate over the appropriate steps to address climate change and the more focused approach with attention to details used to create a cap-and-trade program for power plant emissions of SO_2, NO_x, and mercury. The cabinet-level working group included the secretaries of state, energy, commerce, interior, agriculture, and treasury; the EPA administrator; the chairmen of CEA and CEQ; the director of the National Economic Council; the national security adviser; and the vice president. This group met roughly half a dozen times between July 2001 and February 2002 to discuss climate policy.

Between meetings, CEQ and CEA developed materials and circulated them to various principals. These materials included information on emissions forecasts and historic data, formulation of climate change goals, and descriptions of policies and measures that could be used to achieve those goals. Staff at other agencies, including EPA and the Departments of State, Energy, and Agriculture, also contributed to the development of these materials.

Intensity Targets

Early in the process, the staff began developing ideas for an alternative to absolute emissions targets with a fixed number of tons. The notion of an absolute target, which presumably would only decline over time, is viewed by most economists as undesirable in the near term. Arguments catalyzed by Wigley et al.[44] point out that the development of new technologies in the future, coupled with the natural tendency of capital to turn over gradually, implies that desirable emissions paths will inevitably continue rising before they stabilize and eventually decline in the future. Rather than articulate a goal to limit absolute emissions, which almost sounds like a goal to limit economic growth, the staff considered alternatives that would embrace growth while encouraging emissions reductions below forecast levels.

It was also decided early on to embrace an overall greenhouse gas (GHG) mitigation goal, rather than one of carbon dioxide only. This broad GHG approach emphasized the opportunities to make improvements through reductions in emissions of methane and other industrial greenhouse gases. A broad GHG approach also addressed the suggestions by Hansen et al. and Easterbrook[45] that more expensive reduction in carbon dioxide could be delayed by pursuing cheaper reductions in other gases first, principally methane.

Various articulations of a mitigation goal were researched, including absolute, per capita, growth indexed, intensity, and price-based approaches. An intensity target—emissions per dollar of economic activity—became the consensus choice for a number of reasons. The goal becomes more energy-efficient (or rather GHG-efficient) cars, buildings, and industrial production—without limiting the number of cars and buildings or the amount of production. The goal becomes more energy-efficient transportation, not limits on gasoline use. This was seen as an effective foil for critics arguing that climate policy amounted to limiting economic growth. It was also relatively simple and intuitive. In a world where we are already used to miles per gallon ratings on cars and operating cost per year ratings on appliances, emissions per dollar of gross domestic product (GDP) follows quite naturally. Additionally, it offered a way to show continuing improvement over time—a declining intensity—while emissions could still grow in a way that was consistent with most analysts' views of optimal trajectories.[46]

The notion of an intensity target is not new and, in the end, implies a particular emissions target once GDP is known.[47] Because an absolute emissions target can be designed to increase over time, as implied by a constant intensity target coupled with economic growth, the environmental consequences of an intensity target are not automatically different from an absolute target. Framing the goal can make a difference, however. Environmentalists may like an absolute target approach, which draws a clear line between increases and decreases in actual emissions. Those concerned about economic growth may prefer an intensity target approach, which can achieve real progress without drawing such a line.[48]

Much of the debate comes down to the choice of a particular intensity target, which can be a lot, a little, or somewhere in between. As the working group and support staff considered this question, they studied data produced by the Energy Information Administration (EIA). Each year, EIA produces both a reference forecast and a high-technology forecast that assumes more rapid deployment of energy-saving technologies. The high-technology forecast was an appealing basis for a target because it was grounded in a plausible, but achievable, outcome, in contrast to the Kyoto targets, which were viewed as implausible and arbitrary. Under the December 2001 high-technology forecast, emissions intensity was forecast to fall 16 percent from 2002 to 2012, in contrast to a best-guess reference forecast of 14 percent. With this information, the working group recommended an 18 percent "stretch" goal that went beyond what EIA suggested was achievable with its high-technology assumption. Although 18 percent equals the decline from 1992 to 2002, motivating critics to claim this is nothing better than business as usual, the relevant comparison is with the best available forecast. For any number of reasons—weather, growth, structural change—the next decade is expected to look different from the previous one.

Relative to the reference forecast of a 14 percent decline in intensity, I would argue that an 18 percent intensity improvement is reasonable as an emissions-based goal—that is, a goal looking at emissions levels rather than prices. Part of the problem with setting any kind of emissions goals for greenhouse gas reductions, intensity or absolute, is that a reasonable effort over 10 years could indeed disappear in the noise, as some have argued—not because such a target is inappropriate, but because of the substantial uncertainties arising over a decade. The roughly 4 percent emissions reduction from forecast levels proposed in the Bush Climate Change Initiative is entirely consistent with the average effort of countries participating in the Kyoto Protocol—relative to their business-as-usual forecast—as well as the effort for the United States envisioned by the previous administration under the protocol.[49] It is also similar to those reductions proposed by Senators McCain and Lieberman in the very beginning of their program, taking into account offsets. Most studies of the expected costs of such a policy would peg it around $30 per ton of carbon.[50] Meanwhile, estimates of benefits are centered around $30 per ton.[51]

All of this suggests that such a mitigation goal is reasonable. Problems arise, however, if one looks at the pattern of historic emissions growth. Random 4 percent variation over 10 years is not unusual.[52] An intensity formulation may or may not reduce this variation compared with an absolute target, but either way, a modest target can become twice as hard or become a free ride depending on random emissions shocks—weather, technology, or growth.[53] This suggests that in the long run, a policy that puts more emphasis on stable prices will be necessary, rather than one that emphasizes emissions levels, relative or absolute. One possibility is a cap-and-trade program with a safety valve, whereby an unlimited quantity of additional allowances are provided at fixed price.[54]

Commitment to Market-Based Policies

Although the target is arguably sound, the absence of a strong program to enforce the stated goal is a fair object of criticism. Despite a lively debate over various policy options, a voluntary challenge and tax-incentive approach, coupled with a renewed emphasis on technology, was chosen.[55] As the likelihood of this outcome became clearer, CEA moved away from a position focused on a very modest cap-and-trade program and began looking for institutional improvements that would lead future policy in an economically sensible direction.[56] Such an approach was consistent with the 2002 *Economic Report of the President*, which was being written almost simultaneously with the development of the climate policy. In the report, the theme was institution building; efficient, welfare-improving policies do not spring up overnight, especially in the environmental arena. Rather, there is usually a gradual institutional development: formal rules, markets, monitoring, and administration, as well as informal experience and knowledge. A strong desire existed at CEA to exercise the idea of institution building in the context of laying out a new climate policy.

The two areas where institutional improvements appeared feasible were the development of an emissions registry and the characterization of likely future policies. A greenhouse gas emissions registry already existed at DOE, as established under the 1992 Energy Policy Act. The cabinet-level working group considered how the registry might be changed. One option was to make the registry mandatory and require all sources over a certain emissions threshold to report. This was an option supported by many environmental groups, including Natural Resources Defense Council (NRDC) and the Pew Climate Center.[57] Another option was to use the registry to provide some kind of credit that would be valid against any future requirements, should they arise. Yet another option was to improve the registry's rules, responding to the criticism that it was so flexible as to be meaningless.[58]

CEA's opinion was that a mandatory reporting requirement served little purpose, as adequate aggregate emissions measures were already available based on fuel production data. CEA also felt that should a tradable permit system arise in the future, it should focus on permitting at this same level, at the point of fossil fuel production or distribution, rather than combustion. This would lead to broader coverage while lessening the administrative burden: there are thousands of fossil fuel producers and distributors, but hundreds of millions of consumers.[59]

Mandatory emissions reporting at best suggested that a future tradable permit system would be fashioned at the point of combustion rather than production.[60] Such a downstream system would necessarily exclude small sources, possibly exclude politically vocal sources, and involve far more regulated entities—tens or hundreds of thousands of points of regulation rather than thousands.[61] At worst, it foreshadowed a more onerous regulatory style that would target individual emissions sources with specific and inflexible requirements. The arguments that mandatory reporting offered little improvement in the understanding of aggregate emissions and that it foreshadowed a more onerous regulatory approach in the future were sufficient to dissuade the working group from this choice.

The idea of improving the validity of reported emissions and offering credits, however, arguably leads institution building in the right direction. On the one hand, any future market-based climate policy is likely to require a registry to address voluntary reductions outside the market system. For example, certain emissions sources—fugitive emissions from agriculture and forestry, sequestration, and some international activities—are unlikely to fall under a mandatory program. In order to create incentives and opportunities to reduce such emissions, a registration program of some sort is necessary, and an enhanced voluntary registry could be the genesis of that future program.

In a similar way, a crediting program that provides a homogeneous measure of reduced emissions encourages businesses to think about ways they can reduce emissions. Unlike a program that just registers activities, where later evaluation might distinguish some actions from others, the provision of tradable credits implies that all credits will be treated the same. Further, firms without reduction opportunities can buy credits from others that do. This suggests that a credit market could arise, providing experience and familiarity that again bolster the opportunity for future market-based policies. While the value of a credit is speculative, dependent on expectations of a future trading program and acceptance of the credits, such credits do provide a hedging opportunity for firms, individuals, and even communities that might suffer under future climate policies. Again, these economic arguments held sway and were supported by the working group.

The last area where an opportunity existed to encourage market-based approaches was in the statement about what would happen should current policies be insufficient to meet the president's goals. The president made the following statement in his February 14 speech: "If, however, by 2012, our progress is not sufficient and sound science justifies further action, the United States will respond with additional measures that may include broad-based market programs as well as additional incentives and voluntary measures designed to accelerate technology development and deployment." Considerable debate had surrounded the wording of this statement, and CEA encouraged a strong commitment to market-based programs. The statement that emerged, which mentioned market-based programs first but placed them alongside possible incentive and voluntary measures to encourage new technologies, was admittedly disappointing. A proposed alternative would have more explicitly endorsed a market-based program in the future should the current approach fail. Yet, given the decision that the current situation did not warrant a mandatory program, it would have been difficult to explain why the administration would commit to such a program in the future. That is, how can one argue that a mandatory program is warranted in 10 years only if a voluntary program does not work, without believing a modest mandatory program makes sense now?

Climate Policy in Context

The relatively weak commitment of voluntary programs and tax incentives to meet the climate change goal stands in strong contrast to the enforceable caps proposed under Clear Skies. But is that really the right comparison by which to judge the Climate Change Initiative? Before the 18 percent goal, the United States had never stated a practical domestic target. During the Clinton administration, there were nominal claims to seek a return to 1990 emissions levels, but that was as impractical then as it is now. Both the Jeffords–Lieberman and McCain–Lieberman bills similarly focus on 1990-level emissions, either in 2009 in the electricity sector or by 2015 for most of the economy.[62] Simply by articulating a plausible mitigation goal—one that could eventually be the basis for a feasible cap-and-trade system—the president arguably moved the debate forward. The administration also squarely tackled one of the two vexing problems embodied in the 1997 Byrd–Hagel resolution: namely, how to define a mitigation goal that limited the threat to economic growth.[63]

Was it right to expect more? Recall the circumstances of the 1990 Clean Air Act Amendments: consensus was already building around a 10-million-ton emissions reduction even before President George H.W. Bush proposed it. Consider the context of Clear Skies: the Clean Air Act will require reductions of SO_2, NO_x, and mercury regardless of whether Clear Skies is enacted. On climate change, no groundwork has been laid for a reasonable domestic target: stakeholders on the right and left remain miles apart. In this circumstance, who would support a reasonable mandatory program? Simply articulating the right kind of goal is a development. Meanwhile, the administration also proposed a transferable crediting program that could prove useful as a future trading system evolves.

CONCLUSIONS

As the description of the policy process indicates, economic principles were integral to the development of both the Clear Skies and Climate Change Initiatives. The Clear Skies architecture provides a relatively efficient approach to regulating emissions of SO_2, NO_x, and mercury. Relative to the Acid Rain program, Clear Skies replaces a tangle of NSR and MACT regulations with a simple market-based system, offers improved protection against unexpected cost shocks, provides a gradual transition to an allowance auction, and moves marginal costs more in line with marginal benefits. Various stakeholders may well disagree with the particular allowance allocations or caps envisioned for SO_2, NO_x, and mercury. A casual read of the scientific literature suggests that increased controls of SO_2 could be welfare improving and raises questions about how to handle a pollutant such as mercury, where the benefit estimates are virtually nonexistent. But based on the requirements and likely actions under the Clean Air Act, it is arguably a sensible alternative—improving the environment and reducing economic costs relative to the status quo. The benefits almost certainly outweigh the costs.

It also makes sense to peel off CO_2 and other greenhouse gases into a separate policy. With NO_x, SO_2, and mercury, end-of-pipe treatment is used to reduce emissions, consequences are regional, power plants face some of the lowest control costs, and energy prices are unlikely to rise noticeably. With CO_2, on the other hand, reduction is achieved through conservation and fuel switching, the consequences are global, cheap mitigation opportunities exist beyond the power sector, and market-based policies will need to raise fossil energy prices to be effective. CO_2 mitigation needs to be approached in a broad, economywide context. The Bush climate policy makes headway: it establishes an achievable target and addresses concerns about economic growth through the intensity framework. It also lays institutional groundwork for future policy through the enhanced registry and the transferable credit program.

Where the policy arguably comes up short is in the absence of a clear commitment to broad, market-based controls on greenhouse gases. Billions are being spent on scientific and technological research, billions in tax incentives have been proposed, and some form of greenhouse gas emissions regulation looms on the horizon. In this environment, it would be reasonable to provide more direction and leadership toward an economically sensible policy to reduce emissions—if not to actually propose a particular policy, such as a modest tradable permit system with a safety valve. A cap-and-trade, safety-valve program designed to create a $10 per ton incentive and reduce 30 million metric tons of carbon equivalent emissions might cost the economy $150 million. Alternately, a policy designed to create a $30 per ton incentive and reduce 90 million metric tons— roughly the goal of the Bush climate policy—might cost the economy $1.5 billion. Both programs, however, imply transfers of tens of billions of dollars as permits are distributed, bought, and sold. Although the time may not be ripe for the kind of grand bargain that occurred when the Acid Rain program was conceived, it is fair to ask whether the proposed decade of volunteerism and tax incentives is a sufficient step.

Held to the standard of making progress toward efficient, cost-effective solutions, both policies get check marks. Clear Skies is not perfect, but it is a substantial improvement over implementation of the Clean Air Act. Meanwhile, proposing a sensible mitigation target for greenhouse gases and doing some institutional groundwork are a positive step. Of course, economic principles are not the only metric, and clearly factors other than cost–benefit analysis affect policy choices—as they should. The country is not a technocracy; the sentiments of multitudes of stakeholders matter. In the case of climate change, factors including the relative novelty of greenhouse gas regulation as well as skepticism among some constituencies are important. Moving too fast can arguably lead to setbacks, as the previous administration's experience with the Kyoto Protocol demonstrates.

There are clearly a lot of different ways to view the process and outcome of an environmental policy development exercise. As an economist at CEA, one has the unique opportunity to present and represent an economic view on every question—a view that does not have an overt constituency or a necessarily human bias. At the same time, one gains an immense appreciation for the wide range of other views that filter into a policy decision. As I went off one morning to present and explain the administration's new climate policy to a group of environmentalists, I worried aloud to a more experienced CEA colleague. How could I dutifully represent the administration when I disagreed with key elements of the policy, in some cases based on my own research? The colleague advised me that in similar situations, he would explain the administration's policy; the wide range of concerns that influenced the policy, including his own research; and then that, in the end, the president makes a decision about how to weigh these conflicting concerns, balancing them in a way that he believes makes the most sense for the American people.

Although it may sound a bit corny, I believe there is real truth to that idea. One may not agree with the views of the president and the administration about what makes sense for the American people. But in the end, as an economist at CEA, one has the opportunity to make the case for economic principles and knows that those principles are at least considered, if not followed. One also gains a great appreciation for a vast range of other, noneconomic concerns that must be considered in the decision-making process. Economics should inform decisions, but it cannot dictate them.

NOTES

1. See, e.g., the discussion of the political process leading up to the 1990 Acid Rain trading program in Hausker (1992).

2. The fact that Clear Skies recommends a gradual transition to an allowance auction has not gone unnoticed. See Braine (2003).

3. Senators Tim Wirth and John Heinz released their Project 88 report, calling for increased use of market mechanisms, just prior to debate on the 1990 amendments. See Stavins (1988).

4. For a description of the RECLAIM market failure, see Coy et al. (2001).

5. Speech at G8 Environmental Ministerial Meeting, Trieste, Italy.

6. At the same time the administration was developing its multipollutant bill, alternative proposals were put forward by Senators Jim Jeffords and Joe Lieberman (Senate Bill 556, 107th Congress) and Senators Bob Smith, George Voinovich, and Sam Brownback.

7. Power plants, unlike industrial boilers, are required to report detailed information on plant operations, fuel use, and emissions; unfortunately, data on industrial boilers are more sketchy.

8. This is not true in cases where it is important that allowance trading raise output (electricity) prices and encourage conservation as a significant margin for reducing emissions. Reductions in a three-pollutant setting, excluding carbon dioxide, occur almost entirely through abatement, however, and not through energy conservation. This is in contrast to a four-pollutant setting that includes carbon dioxide, where energy conservation can be responsible for half of total emissions reductions. See Goulder et al. (1999b).

9. See Burtraw and Mansur (1999).

10. Montero (1999) discusses issues surrounding voluntary opt-in in the Acid Rain program.

11. See discussion in Stavins (1998).

12. Not surprisingly, it sometimes seemed like certain stakeholders favored generous opt-in provisions in order to raise the effective caps on power plants, rather than to encourage broader participation and efficiency.

13. Both Stavins (1998) and Ellerman et al. (2000) discuss the selection bias associated with opt-ins under the Acid Rain program.

14. Although this requirement might appear to be well defined, EPA can define source categories in ways that differentiate the requirements on the basis of boiler type or coal rank, recognizing that control options vary. The data on mercury control technology were also limited, adding to the cost uncertainty.

15. The Clean Air Act provides some flexibility for states to meet required air quality standards, such as the new ozone and fine particulate standards developed in 1997. From the time of designation as "nonattainment," states have up to five years to reach the standards, although one five-year and two one-year extensions are possible. The actual deadline, whether less than or greater than five years, must be "as expeditiously as practicable." See Section 172 (a)(2)(A) of the Clean Air Act.

16. The act directs EPA to issue controls on upwind emissions sources if they "significantly contribute" to nonattainment problems in downwind states-authority that the agency first used in September 1998, when it issued the NO_x SIP call addressing interstate patterns.

17. An interesting term I picked up during this debate was "baseball arbitration," referring to a process whereby a mediator will eventually choose one side's position or the other's—and nothing in the middle—at the end of an inconclusive negotiation. The idea is to force the two sides toward a reasonable solution, wherever it lies, as the side with the most unreasonable position is likely to lose.

18. EPA presentation at Edison Electric Institute, September 18, 2001; see January 15, 2002, press release, "Edison Electric Readies Dirty-Air Plan," Clean Air Trust.

19. See U.S. EPA (2002a), Technical Addendum, page 5.

20. See ibid., page 10.

21. See OMB (2000), 46.

22. See U.S. EPA (2002b), C48.

23. Lutter and Burtraw (2003) suggest interpollutant trading as a way to reduce the costs of obtaining the same level of environmental benefits. This idea arose in the working group, but unfortunately too late in the process to be seriously debated. It is unclear, given the specific constraints of the Clean Air Act requirements concerning ozone, whether this would have worked.

24. Lutter and Irwin (2002) provide a summary and sketch a risk assessment suggesting that the public health effects from mercury exposure are small.

25. See U.S. EPA (2002b, C48); the price cap on mercury allowances is $2,187.50 per ounce. Note that cost of mercury controls rises quite rapidly for higher reductions, such as the 90 percent proposed in Senate Bill 556 (the Jeffords–Lieberman bill in the 107th Congress).

26. See U.S. FDA (2001).

27. At the time of this writing, two bills had been proposed as alternatives to Clear Skies in the 108th Congress: Senate Bill 366 (the Jeffords–Lieberman bill from the 107th) and Senate Bill 843 (Carper). Senate Bill 366 has considerably tighter caps and includes CO_2; Senate Bill 843 is in between. See Lankton et al. (2003).

28. See Coy et al. (2001) and U.S. EPA (2002c). Data on RECLAIM market prices from Cantor Fitzgerald.

29. Use of the safety valve must continue each year to maintain the original increase in emissions, and the safety valve price will hold until the original limit is restored. Assuming technologies to further reduce emissions are available at levels below the safety valve price, the environment will eventually be made whole.

30. See Weitzman (1974) and Roberts and Spence (1976).

31. Indeed, discussions about appropriate caps often focused on the location of the "knee" or "elbow" of the marginal cost curve—that is, the level of emissions where marginal costs began to rise more rapidly.

32. Historic variation in the Acid Raid program has been a factor of three: a low of $69 per ton was reached in March 1996, followed by a high of $212 per ton in May 1999. The revised RECLAIM program established an effective safety valve at $15,000 per ton, or about five times the level expected to encourage necessary reductions.

33. This is primarily a concern for pollutants where energy and electricity conservation, as opposed to end-of-pipe treatment by the electricity generator, is a significant component of emissions reductions. It turns out for SO_2, NO_x, and mercury, there is little effect on electricity prices, conservation is minimal, and concerns over inefficiencies created by updating allocations are not immediately relevant except as a precedent for future emissions reductions.

34. Goulder et al. (1999a).

35. Any updating allocation inefficiently encourages fuel switching over conservation by reducing electricity prices. An updating allocation focused on fossil fuels is even worse because it inefficiently encourages too much *among* fossil fuels rather than *between* fossil and nonfossil fuels.

36. E.g., Braine (2003), NMA (2002), Holly (2003).

37. See discussion in U.S. EPA (2002d).

38. This exemption does not cover new sources, sources within 50 miles of a Class I area, or sources that increase their maximum hourly rate (i.e., increase their capacity).

39. An interesting discussion on environmental federalism can be found in Oates and Schwab (1996).

40. Despite the relatively large share of coal combustion in total mercury emissions, the fraction in oxidized form, which is responsible for local effects, is thought to be less than that in other source categories (U.S. EPA 1997c, ES-5).

41. CEC (2002) discusses the reverse auction for renewables.

42. See Johnson (2003).

43. Although highly uncertain, it is not impossible to estimate the benefits of CO_2 mitigation. Estimates are discussed in Pearce et al. (1995) and Smith et al. (2001) and have a mean of around $30/tC for current emissions.

44. Wigley et al. (1996).

45. Hansen et al. (2000) and Easterbrook (2001).

46. Even more recent scenarios (Morita and Robinson 2001, Figure 2.14) similarly indicate that emissions will rise for the next couple of decades for all but the most aggressive stabilization targets.

47. See Pearce et al. (1995). The more flexible idea of a growth-indexed target was adopted by Argentina in 1999 (Secretariat for Natural Resources and Sustainable Development, 1999).

48. This is discussed in Pizer (2003).

49. These two statements often spark considerable debate, but they are nonetheless true. Table 23 in EIA (2000b) indicates that no reductions would be required of the remaining Kyoto participants, as a whole, to meet their overall commitment. Similar forecasts by researchers at MIT indicate a 7 percent reduction (Reilly 2001). One can find details of the domestic reductions required under the Kyoto Protocol, according to the Clinton administration, in an appendix to Janet Yellen's March 8, 1998, testimony before the House Subcommittee on Power and Energy, Committee on Commerce. Specifically, page 323, lines 26 and 29, show that the reductions associated with the quoted cost of $14 per ton correspond to 100 million metric tons of domestic reductions in 2010—almost the same as the Bush administration target.

50. See Weyant and Hill (1999) for likely costs of a 4 percent reduction based on a variety of models; Paltsev et al. (2003) report similar costs for the McCain–Lieberman proposal.

51. See Pearce et al. (1995) and Smith et al. (2001).

52. Historic growth in carbon dioxide emissions has averaged 1 percent over 1980 to 2000, with a standard deviation of 2.4 percent.

53. It turns out that intensity has been less volatile in the United States over the past 20 years. The same is not true for other industrialized countries. See Pizer (2003).

54. See Pizer (2002).

55. The tax incentives—an investment tax credit for combined heat and power facilities, a tax credit for residential solar capacity, extensions of the production tax credit for wind and biomass, a new production tax credit for landfill methane, and a tax credit for hybrid and fuel cell vehicles—were part of the 2001 National Energy Policy and included in the president's 2003 proposed budget. See EOP (2002) and OMB (2002).

56. That CEA was on the losing side of the debate over climate policy was noted by the *Economist* (Lexington 2002).

57. See Pew Center on Global Climate Change (2002).

58. See criticism in Lashof and Hawkins (2000).

59. See Hargrave (1998).

60. This is, in fact, the direction of the recent McCain–Lieberman bill, which specifies regulation of industrial and commercial sources at the point of combustion, excluding entities with emissions below 10,000 metric tons of carbon dioxide. See Pizer and Kopp (2003).

61. The European Union Emissions Trading System, a downstream greenhouse gas emissions trading program, includes 12,000 to 13,000 sources (Kruger and Pizer 2004).

62. McCain–Lieberman is more reasonable in the first phase, based on year 2000 emissions plus an additional 15 percent in offsets.

63. The other problem is the need to obtain significant participation among developing countries.

CHAPTER TWO

HEAD IN THE CLOUDS
DECISION-MAKING:
EPA'S AIR QUALITY
STANDARDS FOR OZONE

Randall Lutter

The 1997 national air quality standards of the Environmental Protection Agency (EPA), potentially the most important environmental regulations of the Clinton administration, have gone through an epic court battle. In May 1999, a panel of the Federal Appeals Court for the D.C. Circuit blocked EPA's ozone standards, stating without dissent, "EPA cannot ignore possible health benefits of ozone."[1] Both plaintiffs and judges cited Lutter and Wolz, who argued that tropospheric or ground-level ozone, like ozone in the stratosphere, reduces human exposure to harmful ultraviolet radiation, and that EPA should consider this quantifiable and potentially significant effect in setting air quality standards.[2] The EPA Administrator Carol Browner responded by declaring that aspect of the court's decision "the most bizarre."[3]

Administrator Browner's polemical condemnation of an uncontested decision by a panel of the Federal Court of Appeals is striking and stands out even among the policy fights that pass for serious debates in Washington. It is educational and enlightening, however, to students of environmental policy, even though it concerns only a narrow pollution issue, the benefits associated with one type of air pollution, ozone. Although in some sense it is history—the Bush administration essentially accepted the Browner response in January 2003—a close reading of the record suggests trouble with the view that the administration's support of the Browner position has validated it and hints at deeper difficulties. The tribulations that EPA has endured in assessing and considering benefits of pollution speak volumes about environmental policymaking in Washington today and about EPA.

Other aspects of the court case are as enlightening. In the second court decision on the ozone standard in 2001, the Supreme Court upheld EPA's practice of setting air quality standards without regard to cost or feasibility.[4] This practice removes any legal constraints that might prevent EPA from issuing standards that are unattainable.

In this chapter, I explain how EPA developed the rule and then offer a critique of the rule. EPA's decision appears to be an overzealous grab for more administrative authority and a willingness to ignore unpleasant facts. Although the Bush administration inherited the rule and the ongoing lawsuits, the new EPA leadership has continued the preexisting policy of neglecting the benefits of ozone. The agency finds it easier to act as if such benefits do not exist than to antagonize the environmentalists who find the mention of these benefits heretical.

My experience and my research have led me to two broad conclusions. First, EPA's ozone standard set a low in the use of bad analysis to support bad environmental policy.

- EPA ignored estimates of the lost benefits of low-level ozone that were prepared by its own contractor, even when these appeared similar in magnitude to the health benefits EPA used to justify the standard.
- In estimating the cost of the rule, EPA deliberately and inappropriately assumed that the necessary emissions reductions would never cost more than an arbitrary amount, even in cities where these reductions were so large they could not be achieved by the elimination of motor vehicle emissions. Many cities today appear to have trouble meeting the preexisting standard by congressionally imposed deadlines.
- Meeting the 1997 standard appears to be infeasible and to involve costs many times greater than EPA's most optimistic estimate of the benefits.

EPA kept these facts from the public by selectively disseminating its analyses.

Second, the procedures intended to guarantee the quality of the risk assessment and full reporting of costs and benefits failed. Lasting improvements in the assessment of environmental risks, costs, and benefits will come only with greater involvement of parties independent of the executive branch.

I became involved with EPA's national ambient air quality standards (NAAQS) for ozone in the context of regulatory oversight efforts, which several administrations have formalized within the Executive Office of the President to help rationalize regulatory policy. As a staff economist for the Office of Information and Regulatory Affairs (OIRA) at the Office of Management and Budget (OMB), I reviewed draft regulations that agencies wanted to propose or to issue. Along with other OIRA staff, I assessed whether agencies' benefit–cost analyses were consistent with best practices and, based on the analyses, whether draft regulations were consistent with the economic principles enumerated in presidential executive orders dealing with regulations. President Clinton's Executive Order No. 12866 on Regulatory Planning and Review, for example, stipulates that benefits should justify costs, regulations should achieve stated regulatory goals cost-effectively, and agencies should regulate only when necessary based on an assessment of costs and benefits.

Staff analysts from OMB and other agencies had a unique opportunity to participate in the development of the air quality standards that EPA issued in 1997. Staff from OMB and other agencies typically reviewed regulatory proposals and final rules in draft form only during a period of up to 90 days prior to public release of rules. In 1993, however, incoming Clinton administration officials sought to involve OIRA staff in early stages of a few selected regulatory initiatives, hoping that this would improve the often antagonistic relations that then existed between staff from OMB and EPA. OMB and EPA officials agreed that the air quality standards would be one initiative where OMB staff would meet early and regularly with EPA staff about the development of new standards. As a result, from 1994 to 1997, analysts from OMB and other agencies met frequently with EPA staff about EPA's efforts to revise the national ambient air quality standards.

NEGLECTING OZONE BENEFITS

A simple question in an interagency meeting in fall 1994 sparked a big dispute. Staff from OMB and CEA asked EPA whether ozone near the ground—tropospheric ozone—like stratospheric ozone, could absorb solar radiation and so reduce risks of skin cancers and cataracts associated with too much sun. The question's genesis was a discussion with a former CEA economist who had previously worked on both stratospheric ozone and tropospheric ozone programs and understood that total ozone determines the level of screening. In addition, OMB staff, having reviewed EPA's earlier bans on substances that reduced stratospheric ozone, were aware of both the benefits of ozone and EPA's risk-assessment techniques.

Underlying the original question about the health benefits of ground-level ozone was an incipient frustration among OMB and CEA staff with the way that EPA said it would arrive at a health-based standard. The Clean Air Act directed EPA to set a standard to "protect public health" with an "adequate margin of safety." In earlier rulemakings, EPA had interpreted this language to preclude *any* consideration of cost or feasibility, and appellate courts had concurred that this interpretation was permissible. As a result, in setting the standard, EPA planned to consider only the respiratory health benefits of lower ozone concentrations. Such an approach may have made sense given earlier scientific views that there was a threshold in the relationship between concentration, or dose, and health effects. EPA, if it could identify such a threshold, could set a health-based standard in a transparently rational manner.

New evidence that the relationship between ozone concentrations and health effects was linear made EPA's traditional approach difficult to understand; indeed, it seemed largely subjective. EPA planned to choose how clean was clean enough by considering the totality of the evidence of respiratory health effects and making a judgment. Since the respiratory health effects of ozone concentrations fall (and the evidence for such effects diminishes) as ozone concentrations fall, the judgment appeared largely arbitrary. There was little if any role for economic analysis. In particular, EPA's interpretation of the Clean Air Act appeared to give it nearly unbridled discretion to set the standard that it thought best. This discretion was at the heart of a controversial finding by the D.C. Circuit in the *American Trucking* decision of May 1999 that EPA's ozone standard represented an unconstitutional delegation of lawmaking authority to the executive branch.

In a second 1994 meeting, EPA staff responded to the original OMB and CEA request by presenting estimates indicating that a decline in seasonal average concentrations of low-level ozone of about 10 parts per billion (ppb) would lead to a change of 0.5 percent change in total column ozone.

Since such an estimate cannot usefully be judged either large or small, OMB and CEA staff asked for an assessment of the health risks associated with such a change. EPA staff later presented an estimate that the health risks of the decline in ozone concentrations were about 300 cancers per year. But in a follow-up meeting, they refused to share the analysis that led to this estimate. They also declined to conduct any further analysis of this issue, claiming that they were acting at the direction of their management.

Staff analysts at OMB, frustrated that EPA had refused to estimate adverse health effects of a proposed government action, contacted the Department of Energy (DOE) to ask analysts there to assess the health effects from increased exposure to harmful ultraviolet radiation (UV-B). DOE scientists used the EPA estimate that a 10 ppb decline in seasonal average ozone concentrations amounts to a 0.5 percent decline in the total column of ozone, which is relevant for predicting changes in UV-B. They presented their results at a public meeting of EPA's Clean Air Science Advisory Council (CASAC), the body charged by the Clean Air Act to review EPA's assessment of health risks.[5] They estimated that a 10 ppb change, if observed nationwide, would lead to annual increases of 2,000 to 11,000 nonmelanoma skin cancers, 130 to 160 melanomas, 25 to 50 deaths, and 13,000 to 28,000 cataracts. DOE staff recommended that EPA consider these effects in setting the health-based standard, noting that they were generally as well understood as the respiratory health effects that EPA used to justify its standard.

Some members of CASAC, who were generally chosen for their expertise in respiratory matters and related issues of atmospheric chemistry, did not understand the implications of the risk–risk tradeoff that DOE staff presented to them.[6] One member who had worked on stratospheric ozone problems then posed the key question: "Should this group [CASAC] take responsibility for this problem?"[7] EPA staff replied, "We do not think . . . that the health effects of UV-B have to be added to the Criteria Document."[8] EPA staff also noted that there was as yet no decision on the separate policy question of whether UV-B health effects should be used to set the standards.[9] Based on this advice from EPA staff, the 1996 Criteria Document excluded the DOE's staff assessment of UV-related health risks and CASAC's views about how to incorporate such risk estimates into the standard-setting process.

One member of CASAC, Dr. Warner North, offered a revealing insight a few months later. He allowed that CASAC may have dropped the ball if (as is true) the Clean Air Act stipulated that the risk assessment that CASAC was reviewing had to include all "identifiable"—as opposed to all adverse—effects. Yet the primary responsibility for the exclusion of any assessment of benefits of ozone from EPA's risk assessment lies with EPA staff who knew but did not mention that CASAC was officially charged with assessing all "identifiable" effects.

Building in part from DOE's presentation, I developed with Chris Wolz, another OMB analyst, an assessment of the health benefits from low-level ozone and a survey of the scientific evidence in favor of such benefits.[10] Wolz and I argued that these benefits were identifiable, which under the Clean Air Act meant they had to be included in EPA's risk assessment. All models of UV-B flux stress that ozone at all altitudes, not just in the stratosphere, acts to reduce ultraviolet radiation, and field measurements show that low-level ozone measurably reduces UV-B. Long-standing concerns about the health effects of UV-B motivated EPA's successful effort to ban chemicals that reduce stratospheric ozone. Thus, the relevant question is the magnitude of the UV-B related health effects. On the basis of a risk assessment that we developed to corroborate the DOE estimates, we predicted that 4,200 to 8,100 cases of nonmelanoma skin cancer per year would result from the 10 ppb decline in seasonal average ozone levels.

More important, we sought to develop a common metric to compare the adverse effects of ozone with its beneficial effects. In particular, we used EPA's valuation methods to assign dollar amounts to the UV-B health effects. We concluded that these would total between $0.3 billion and $1.1 billion per year—amounts as large as the value of respiratory health benefits that EPA later used to justify its rule. We outlined an analytic approach that would let EPA set a standard to maximize public health given the risk–risk tradeoff implicit in any change to tropospheric ozone concentrations. An important unresolved uncertainty, however, was whether attainment of the standard would indeed reduce seasonal average concentrations by 10 ppb.

We got little support from government analysts outside OMB when we sought comments on our draft paper. Staff at DOE said that they had received instructions not to do any work on the issue. EPA staff simply did not respond to our informal draft. We submitted our manuscript for publication with *Environmental Science & Technology.*

In the meantime, Dr. Larry Cupitt, an atmospheric scientist with EPA, estimated that a 10 ppb reduction in seasonal average ozone concentrations would lead to 3,000 to 4,000 cases of nonmelanoma skin cancers per year. A subsequent EPA memo critiqued these estimates, claiming they "cannot be defended." That memo failed to argue, however, that they were too high and failed to show that they were any more uncertain than EPA's estimates of respiratory health effects.

When the Lutter and Wolz paper appeared in *Environmental Science & Technology,* several industry organizations immediately submitted it as part of their public comments to the EPA docket. Chris Wolz and I were then "recused" from interagency policy deliberations related to the health benefits of low-level ozone. Such treatment is typically reserved for policymakers who have conflicts of interest that jeopardize the public appearance of impartiality.

In July 1997, an analysis arrived at OMB from EPA, which stated, "The methodology for estimating such increases [of both UV levels and skin cancer incidence] is *well established* [emphasis added]."[11] It predicted that 696 additional nonmelanoma skin cancers per year would result from the air quality improvements expected if the standard were met. The draft analysis went beyond all previous work by using the distribution of air quality changes expected to occur as a result of the new standard, instead of a uniform decline of 10 ppb in seasonal average ozone concentrations. EPA acknowledged in the final response to the remand that this memo was written by Dr. Sasha Madronich, an EPA contractor.

The draft 1997 analysis has, however, two shortcomings. First, it assumed that the entire population within each of the forty-eight contiguous states would experience the change in air quality projected to occur on average throughout the state. In fact, the United States is very urbanized, and most of the population resides near relatively polluted cities such as New York, Chicago, and Los Angeles, where the declines in ozone concentrations will be much greater than state averages. As a result, the estimate is biased downward.

Second, the 1997 Madronich analysis used as a baseline *current* air quality rather than the air quality consistent with full attainment of the standard that was in place at the time. Thus, the air quality from which the Madronich analysis assesses the effects of a new standard is more polluted than it should be and the change in air quality greater than appropriate. This assumption implies that the estimate of 696 nonmelanoma skin cancers is larger than from a baseline of full attainment of the current standard; thus, the estimate is biased upward.

The 1997 analysis assessed only some of the UV-B related health effects— those related to nonmelanoma skin cancers. One can derive a back-of-the-envelope estimate of the other health effects by assuming the estimates included in the 1997 analysis have the same ratios to nonmelanoma skin cancers as reported in the earlier DOE study. In this case, the total annual number of additional cases would range between 10 and 46 melanomas, 3 and 9 melanoma fatalities, and 1,800 and 4,600 cataracts.

After the rule was issued in July 1997, several industry groups sued, and in May 1999, the Court of Appeals for the D.C. Circuit found that UV-B related health effects are identifiable under the Clean Air Act and therefore must be considered by EPA. It wrote, "It seems bizarre that a statute intended to improve human health would, as EPA claimed at argument, lock the agency into looking at only one half of a substance's health effects in determining the maximum level for that substance."[12] It remanded to EPA "to formulate adequate decision criteria for its ordinary object of analysis—ill effects. We leave it to the agency on remand to determine whether, using the same approach as it does for those, tropospheric ozone has a beneficent effect, and if so, then to assess ozone's net adverse health effect by whatever criteria it adopts."[13] EPA did not contest this decision when it successfully appealed the rest of the Appeals Court ruling to the U.S. Supreme Court.

At the end of the Clinton administration in January 2001, then–EPA administrator Browner signed a proposed response to the Court of Appeals remand.[14] The response stated that information linking changes in ground-level ozone concentrations to changes in exposures to UV-B radiation was too uncertain to warrant relaxing the standard issued in 1997. It also said that associated changes in UV-B radiation exposures would likely be very small from a public health perspective. It then embraced the seemingly inconsistent conclusions that the benefits of tropospheric ozone effects are both "unquantifiable" and "small."[15] Finally, it reproposed an ozone standard identical to the one promulgated in 1997.[16] This proposed response to the remand was not published before the Clinton administration left office. Along with many other such pending regulations, its publication was held up until the presidential appointees could review it.

In November 2001, EPA published in the Federal Register a new response to the court's decision that adopted most of the ideas in Browner's proposed response,[17] claiming that EPA had considered the benefits of ozone in setting ozone standards to protect public health.[18] Borrowing the Browner rationale for why it again proposed the standards first issued in 1997, it proposed to find that the benefits of ozone to health were too uncertain to quantify and besides too small to matter. EPA's proposal neglected established procedures to safeguard careful science.[19] In particular, in developing the proposal, EPA did not seek any advice from its Clean Air Scientific Advisory Committee, although both the Clean Air Act and EPA's practices call for this committee to advise EPA. In addition, EPA did not amend its criteria document, which fails to mention the health benefits of ozone. Yet the act requires that EPA's criteria "accurately reflect the latest scientific knowledge useful in indicating the kind and extent of all identifiable effects."[20] The Court of Appeals also ruled that EPA must consider ozone benefits "in formulating air quality criteria."[21] The 2001 proposed response to the remand ignored these directives.

Most surprising, EPA's proposed response, even though prepared under new political management, did not mention the 1997 draft analysis by Madronich, which had concluded that the methodology for estimating increases of both UV-B levels and skin cancer incidence was "well established."[22] In essence, it solicited comment from the public about its proposed response to the remand while neglecting to inform the public about a key analysis that it had let languish for four years. Ironically, in the final rule that it published in January 2003, EPA explained that it could not consider this analysis more fully in developing its response to the remand because the "draft analysis was never completed, published or peer-reviewed."[23] EPA's final rule did not explain why during the five years of public debate about UV-B health effects, it had taken no steps to complete or publish the Madronich memo or submit it for peer review. One is tempted to conclude that EPA was unwilling to acknowledge existence of the memo rule until it had no alternative, and that it never finalized or formally published the analysis because its conclusions undermined the 1997 standard.

The flaws in EPA's response to the remand were not merely procedural. The Madronich memo, which was not mentioned in the 2001 proposal, suggests that the 1997 ozone standards that EPA reproposed in 2001 may entail lost health benefits of ozone roughly comparable to the health benefits of the standards. For nine cities, EPA's calculations suggest that the revised standard would reduce the number of "outdoor children" who experience temporary and reversible episodes of moderate to severe chest pain upon deep inspiration (PDI) by between 10,000 and 40,000 during the warm months of the "ozone season."[24] In addition, EPA estimated that in New York, the number of hospitalizations among asthmatics would fall by about 0.3 percent, or 40 admissions, during the ozone season.[25] Finally, it projected that its standard would reduce by about 200,000 the number of outdoor children experiencing "exposures of concern," potentially leading to nonspecific bronchial responsiveness, decreased pulmonary defense mechanisms, and indicators of pulmonary inflammation.[26] EPA believes that these estimates of respiratory health effects represent only the measurable peak of a pyramid.[27]

Regardless of the possibility of such a pyramid, a heuristic comparison of these generally temporary and reversible respiratory effects with the relatively persistent UV-B related health benefits that would be lost as a result of a more stringent standard illustrates the problem with EPA's assertion that UV-B related health benefits are very small. Hundreds of non-melanoma skin cancers, dozens of melanomas, a number of fatalities, and thousands of cataracts are clearly significant compared with the tens of thousands of cases of temporary PDI, hundreds of thousands of temporary exposures of concern, and dozens of projected hospitalizations that EPA quantified in support of its revised standard. The lost UV-B related health benefits of ozone are comparable, in terms of adverse effect on public health, with the estimates of respiratory benefits that EPA used to justify its standard. Thus, EPA's conclusion that the health benefits of ozone are essentially negligible appears incompatible with the Madronich analysis that the agency commissioned.[28] In responding to public comments, EPA adopted a procedural defense, saying, "In summary, the Madronich draft analysis does not represent the type of peer-reviewed information that is appropriately relied upon as a basis for NAAQS rule-making."[29]

A valuable extension to the Madronich analysis would be to quantify the net health benefits of EPA's 1997 standard by expressing these varied effects on public health in a single metric. Lutter and Wolz (1997) pointed out the importance of such a metric in judging the risk–risk tradeoff posed by ozone. Moreover, the Court of Appeals ruled that EPA must consider "ozone's net adverse health effect,"[30] a concept that presumes the existence and use of such a metric. The medical community has extensively used quality-adjusted life-years in such instances. An alternative metric, willingness to pay, is nearly ubiquitous in the evaluation of regulatory effects because of presidential directives such as Executive Order No. 12866.[31] EPA can and should pursue the measurement of net public health effects in a single metric, in accordance with the advice of its CASAC and the court. It declined to do so in its 2003 final response to the remand because it believed that the UV-B related health effects could not be reliably quantified.

Why does EPA neglect the adverse health effects of its rule, given the legal and policy reasons for it to consider such effects? A simple and revealing answer is because the politics of such neglect are attractive. EPA's decision placates environmentalists, many of whom find abhorrent the notion that pollution could ever have any benefits. A more nuanced answer is that control of EPA has changed little since the early 1990s, despite the new political leadership. The senior civil servants in charge of the Office of Air are largely the same individuals, and the institutional constraints they face are little changed.

PRETENDING THE OZONE STANDARD CAN BE MET

EPA underestimated the costs of meeting the standard in a way that parallels its neglect of the benefits of low-level ozone. In 1995, EPA staff asked OMB staff how to estimate the costs of meeting a new ozone standard so stringent it could not be met even with the adoption of all identifiable emissions control measures. A simple response would be that such a standard is infeasible with current technology. Although any air quality standard is feasible with mandatory relocation of resident populations, political and constitutional constraints on such relocations preclude their use. Thus, if a standard cannot be met after the adoption of all known control technologies, then it should be considered infeasible for all practical purposes.

Given EPA's natural reluctance to declare its new standard infeasible, I suggested to EPA staff that the agency estimate costs by using an extrapolation procedure. EPA could use regression analysis to estimate a cost curve consistent with identifiable emissions control measures. EPA could then extrapolate the cost curve as far as necessary to estimate the cost of the necessary emissions reductions. Since there might be substantial technological change between the time the rule would be issued and the deadline for compliance, around 2010, I suggested EPA use an assumed rate of technological progress to lower the cost curve. Extrapolation far beyond the range of available data, however, leads to great uncertainty.

The analytic problem of extrapolation is ubiquitous in assessments of risks from environmental hazards. Toxicologists and epidemiologists typically estimate an association between the risk of disease or injury and exposure to a hazard at some level of exposure. They then extrapolate this association to estimate risk at much lower doses or levels of exposure. This extrapolation makes such estimates controversial.[32]

In fall 1996, EPA submitted to OMB a draft proposed rule that neither acknowledged the infeasibility of the standard in some cities nor provided a numerical estimate of the costs of meeting the standard. The rule presented cost data only for identifiable engineering control measures, even though these would leave major cities out of compliance. Many observers in government and industry paid little attention to this omission, because the relevant sections of the Clean Air Act, as interpreted by EPA, placed costs out-of-bounds in discussions about what the standard should be.

In December 1997, CEA staff assessed the costs of EPA's ozone standard using a method very similar to the one that EPA staff and I had agreed to years earlier. The result, $62 billion per year, appeared in a draft memo that the administration eventually sent to Congress in response to a request for documents. After this estimate was made public, EPA felt pressure to develop its own estimate.

When EPA's final ozone rule arrived at OMB in early summer 1997, it included an estimate of the full costs of meeting the ozone standard. But instead of developing an estimate based on an extrapolation of the cost curves implicit in the identifiable control measures, EPA arbitrarily assumed that emissions reductions from any unidentifiable measures would cost $10,000 per ton of emissions reduced. This assumption lacked any empirical foundation whatsoever. Shogren later criticized it for being too low by an amount he dubbed the "lost triangle."[33]

When EPA's draft final rule was under review at OMB, agencies began a wrenching process of negotiating official estimates of costs and benefits of the standards, and in particular, the particulate matter standards. EPA's ozone standard, with full attainment costs estimated by EPA to be nearly $10 billion per year, was already among the most expensive environmental regulations of the Clinton administration. (The 1997 fine particulate matter standard, which the administration issued simultaneously, would cost about $38 billion in 2010, according to EPA.) EPA's benefit estimates ranged from $1.5 billion to $8.5 billion. In part because the cost estimate was already high, and in part because the net benefits were already negative, there was little support for a higher cost estimate during the process of agreeing on final cost and benefit estimates.

I later independently reassessed EPA's cost data, which include information on cost and emissions reductions for a set of engineering measures that reduce emissions.[34] In Figure 2-1, I present EPA's data on the cost of controlling volatile organic compounds emissions in Los Angeles. The baseline for the emissions reductions in Figure 2-1 is a scenario for 2010 in which economic growth lifts emissions above current levels but more stringent control measures adopted before 2010 limit emissions. EPA presents emissions reductions measured in tons per day during the ozone season. I present in Figure 2-1 a marginal cost curve fit to these points.[35] I estimate the marginal

Figure 2-1. The Marginal Cost of Reducing Volatile Organic Compound (VOC) Emissions in Los Angeles in 2010

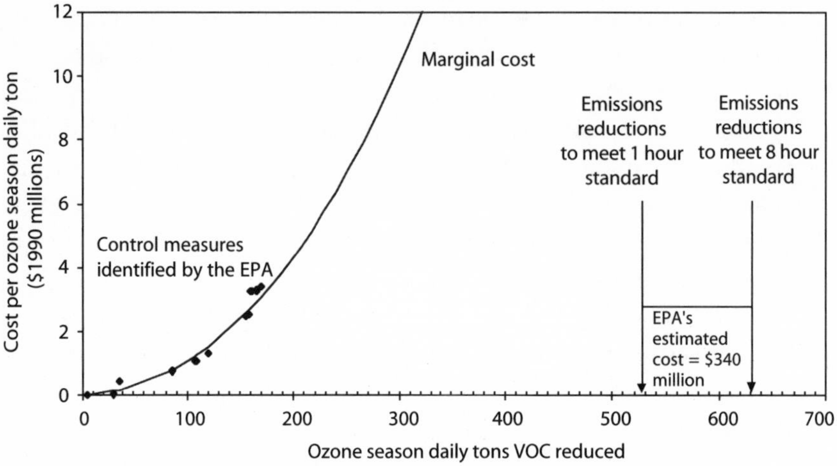

Note: The scatterplot excludes controls that reduce both NO_x and VOCs and controls that have zero cost according to the EPA. The area described as Los Angeles includes Riverside, Orange, and Los Angeles counties. Emissions reductions are from EPA's baseline of 1,054 tons per day. A cost per daily ton of VOCs in Los Angeles of $3.3 million is equivalent to a cost per annual ton of about $10,000. The benefits of controlling ozone are much less than $3.3 million per daily ton.

cost curve by assuming that the log of the incremental cost per ton of identified measures is related to the log of the incremental emissions reductions, so as to ensure normality in the error terms. I use a relationship that is quadratic in logs because this relationship better explains the data.[36]

The total cost of meeting the new standard is the area under the marginal cost curve between two levels of emissions. The first level corresponds to compliance with the old 1-hour standard issued in 1979, and the second level reflects attainment of the 8-hour standard issued in 1997.[37] To estimate marginal cost, EPA simply "assumes the average cost of [emissions] reductions achieved through . . . unspecified methods is $10,000 per ton."[38] Figure 2-1 presents a graphical depiction of EPA's cost estimates as a rectangle.[39] The total cost of reducing volatile organic compound (VOC) emissions to meet the 8-hour ozone standard in Los Angeles in 2010, from a baseline of attainment of the 1-hour standard, is about $340 million per year, according to EPA.

For NO_x, two estimates of the marginal cost of reducing emissions appear in Figure 2-2. Curve *a* is the marginal cost derived from a total cost regression that is cubic in the log of emissions reductions. Curve *b*, derived as for VOCs, is from the regression described in detail by Lutter.[40] Interestingly, curve *b* appears not to fit the most expensive data points. Curve *a*,

Figure 2-2. The Marginal Cost of Reducing NO$_X$ Emissions in Los Angeles in 2010

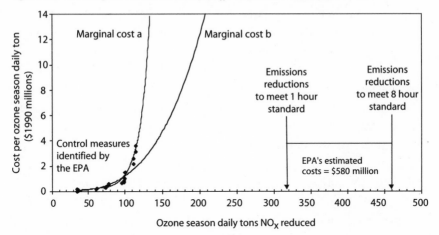

Note: The scatterplot excludes controls that reduce both NO$_X$ and VOCs and controls that have zero cost according to the EPA. The area described as Los Angeles includes Riverside, Orange, and Los Angeles counties. Emissions reductions are from EPA's baseline of 1,048 tons per day. A cost per daily ton of NO$_X$ in Los Angeles of $4.1 million is equivalent to a cost per annual ton of about $10,000. The benefits of controlling ozone are much less than $4 million per daily ton.

which appears to fit the Los Angeles data better than cost curve *b*, implies much higher costs, but it does not fit data for other cities as well as the marginal cost function *b* that I choose to emphasize. EPA estimated the cost of reducing NO$_X$ in Los Angeles to be only $580 million per year, an amount that is much less than the area under either cost curve *a* or *b*.

Technological change would substantially lower the cost estimates implicit in Figures 2-1 and 2-2, but forecasting technology 10 years into the future is extremely difficult. For simplicity, I account for future technological change by estimating the rate of decline in the cost of emissions controls based on the set of new technologies listed by EPA in its regulatory analysis. The arithmetic mean of the average rates of annual cost decline observed among the technologies cited by EPA is 7.7 percent.[41] An annual rate of change of 7.7 percent implies that cost in 2010, 13 years after EPA's analysis was completed, would be about 37 percent of its 1997 value. This estimate overstates likely technological progress, however, because it reflects cost declines only in successful new technologies. Some new technologies, such as nuclear power generation, are adopted but then turn out to be more costly than originally anticipated. To account for such failures among new technologies, I use 5 percent as an average rate of cost decline. In this case, cost would equal 52 percent of the original values by 2010. The cost of meeting EPA's ozone standard in Los Angeles in 2010, based on rates of decline in abatement cost of 7.7 percent and 5 percent, is $8.1 billion and $11.5 billion, respectively.

A recent scientific study indicates that the emissions cuts necessary to meet the standards may be much larger than implied by EPA's analyses. In particular, in Los Angeles, emissions cuts as large as 90 percent may be inadequate to meet EPA's ozone standards.[42]

For other cities, the cost estimates are about $4,700 billion in 2010, assuming that technological progress between 1997 and 2010 will lower cost by approximately half. All but $70 billion of the annual cost occurs in Fresno, California, where EPA estimates that meeting the standard requires cuts in NO_x and VOCs of more than 60 percent. The costs of meeting the standard in Fresno are so high because the emissions cuts in Fresno are much greater than in other cities. For example, to meet EPA's standard, nitrogen oxides must fall by 62 percent in Fresno, but "only" 33 percent in Los Angeles.

Ignoring Fresno, Winner and Cass's (2000) cost estimates for the other cities considered by the EPA, after netting out improvements due to technological progress, are seven times greater than EPA's own cost estimate. These estimates may be too low, because they exclude cities such as Houston and Galveston, Texas, where regulators have had trouble identifying control measures capable of meeting the *old* standard.[43]

These estimates should clearly be viewed with substantial caution, because they are based on extrapolations far beyond the range of available data. They are sensitive to alternative assumptions about the form of the relationship between cost and emissions reductions, although other regressions tend to give similar conclusions.[44] Uncertainties about the functional form most appropriate for such large extrapolations and about the rate of technological progress make quixotic any quest for precise cost estimates. But this should not obscure the fact that EPA's estimates are too low. Moreover, attaining EPA's 1997 ozone standard is infeasible for all practical purposes, something that should be apparent simply from the very large-percentage emissions cuts required in certain cities.

The incremental benefits of meeting the standard in these cities are much less than the costs. EPA estimated benefits of the ozone standard, including environmental benefits unrelated to health, would range from $1.5 billion to $8.5 billion in 2010.[45] The benefits of attainment from a baseline in which all available control measures are implemented would be between $1.1 billion and $6.4 billion in 2010.[46] The emissions reductions necessary to realize such benefits are 1,030 ozone season daily tons of VOCs and 1,700 ozone season daily tons of NO_x.[47] Therefore, the benefits are approximately $0.4 million to $2.3 million per ozone season daily ton of emissions that contribute to ozone formation. Both of these estimates are below the incremental costs of meeting the ozone standard.

These high cost estimates are not meant to suggest that such sums will ever be spent on ozone control efforts. State and local governments will obtain waivers and exemptions from the courts and Congress to keep ozone control costs below the estimates of tens of billions of dollars per year necessary for attainment. The congressionally mandated attainment dates are contingent on an EPA action designating certain areas to be out of attainment, and understandably, EPA has not been eager to make such designations. The practical result of EPA's rule has simply been to give EPA tremendous latitude in requiring emissions cuts in polluted areas.

How did such an obvious error in estimating costs survive the review process?

- First, EPA staff developed the plateau cost curve because it satisfied the political needs of EPA management. They did this knowing that their estimates understated likely costs and had no analytic basis. Analysts from OMB, CEA, and other agencies repeatedly made this point and were ignored.
- Second, CEA, OMB, and other agencies could not compel more accurate cost estimates because there was no constituency in favor of significantly higher cost estimates.
- Third, EPA avoided any public comment on these cost estimates by simply delaying their release until after the public comment period had closed. Such delay was permissible under the Administrative Procedures Act because the cost estimates had no bearing on EPA's ozone standards.

The court and the litigants in *American Trucking* ignored the cost estimates because the law requires that air quality standards be based exclusively on public health considerations. Questions of cost and feasibility were not admissible, and the plaintiffs therefore did not consider them. Although there has been exceptional publicity around EPA's ozone standard, few observers have noticed that Shogren's "lost triangle" is missing.

MAKING BETTER REGULATORY DECISIONS

EPA's analytic errors are not inadvertent. They are the result of efforts to convince the public that the rule was reasonable when the facts indicated otherwise. EPA manipulated its scientific advisers and the public review process. It sought to limit the release of staff analyses with results contrary to the political views and needs of management. It neglected scientific evidence developed by its own risk assessors and by DOE and OMB staff despite Clean Air Act language directing EPA to set standards to protect public health based on "all identifiable effects" of pollutants. EPA knowingly understated costs by ignoring the fact that the incremental costs of emissions reductions rise as emissions reductions grow.

EPA's analyses are distorted, but the ability of motivated bureaucracies to abuse economics to sell pet projects is not news. Environmentalists for years have complained about the Army Corps of Engineers' use of distorted and faulty economics to justify dikes and dams. The *Washington Post* reported that corps staff has complained of such distortions and even sought whistleblower protection.[48] Reforms in this arena are nearly imperceptible.[49] EPA analyses are not qualitatively different.

Can improved analysis contribute to better regulatory decisions in such an environment? It is important to realize that improving the analysis conducted by EPA staff will not necessarily help the agency regulate the environment in a manner that more rationally balances costs and benefits. EPA's actions in developing the ozone standard indicate that it cared little about the tradeoffs important to economists. Administrator Browner stated at an AEI-Brookings Joint Center conference: "To suggest that because you'll have fewer cancer cases with darkened, polluted skies, you should allow for that? That we should allow for more asthma attacks to prevent cancer? I don't think that is what the American people or the Congress had in mind when they passed the original Clean Air Act."[50] On other occasions, Browner has said that no tradeoffs between environmental amenities and economic activity exist. Why expect her to consider tradeoffs that she believes do not exist?

EPA's neglect of tradeoffs between environmental quality and economic costs does not mean that its management is intellectually or morally deficient. A less judgmental and more accurate interpretation is simply that the competitive political world in which EPA operates provides no incentives for EPA to consider seriously such tradeoffs in its decision making.

One road to regulatory decisions that would better balance costs and benefits would be to amend underlying statutes such as the Clean Air Act so that they direct regulatory agencies to consider more fully costs and benefits in their rulemakings. In the last decade, Congress has taken no meaningful steps toward such reform, and it is currently not under discussion.

A more modest approach would be to improve analysis so that it can better inform the public about the expected consequences of regulations. Getting better information to the public depends on the involvement of parties other than EPA. A congressional agency charged with replicating quantitative estimates of costs and benefits developed by regulatory agencies could serve to limit misleading statements about regulatory effects.[51] The OMB's recent policy about peer review and reproducibility of information disseminated by federal agencies is a good step toward improving the quality of information about regulatory effects, but it is too early to say whether it will be sufficient.[52] Replication efforts would provide institutional incentives for regulatory agencies to tell the public what they know about the effects of their actions.

CONCLUSIONS

Integrity in regulatory analysis is important, and the administration should take a strong stand to protect it. In particular, it should support independent calculation including replication of the estimates of regulatory costs and benefits by parties outside the executive branch. Its new policy to improve the quality of information disclosed is a modest step in that direction.[53]

Cost should not be a dirty word in environmental policy. Yet that is the case when Congress and the courts contend that EPA should set air quality standards without any regard to the cost of achieving them. This practice will lead to regulations that are too difficult to be complied with and whose costs bear no relationship to public health or environmental benefits. It also encourages politicians to justify their decisions without regard to cost, a practice that sidesteps public accountability. The administration should advocate reform to environmental statutes so that both costs and benefits are considered in environmental regulatory decisions. Such a policy would promote better regulatory decisions.

NOTES

1. See U.S. Court of Appeals for the District of Columbia Circuit (1999).
2. See Lutter and Wolz (1997).
3. See Kennedy (2000).

4. It also found that EPA must set standards "not lower or higher than is necessary" to protect the public health with an adequate margin of safety. See U.S. Supreme Court (2001, *15*).

5. See U.S. DOE (1995). Lutter and Wolz (1997) also describe the DOE results.

6. See U.S. EPA (1995, *211–219*).

7. Ibid. (*212*).

8. Ibid. Criteria must "accurately reflect the latest scientific knowledge useful in indicating the kind and extent of all identifiable effects on public health or welfare which may be expected from the presence of such pollutant in the ambient air." See Clean Air Act, Section 108(a)(2).

9. See U.S. EPA (1995, *212*).

10. See Lutter and Wolz (1997).

11. This memo appears in the OMB docket for the ozone standard but not in the EPA docket that constitutes the legal record examined by the court. As a result, it was not considered in *American Trucking*. The AEI-Brookings Joint Center for Regulatory Studies has posted the memo on its website at http://aei-brookings.org/admin/pdffiles/php9v.pdf.

12. See U.S. Court of Appeals for the District of Columbia Circuit (1999, *1027*).

13. Ibid.

14. See U.S. EPA (2001a).

15. Ibid.

16. Ibid.

17. Ibid. (2001b).

18. Ibid.

19. See Lutter and Gruenspecht (2001) for a more detailed discussion.

20. See Clean Air Act, 42 U.S.C. Section 7409(b)(1)(1994).

21. See *American Trucking Associations v. the Environmental Protection Agency*, 175 F.3d 1027, 1025 (DC Cir 1999).

22. The AEI Brookings Joint Center for Regulatory Studies posted this analysis on its website in early 2002.

23. See U.S. EPA (2003, *634*).

24. See Whitfield (1997, Table 2). This range reflects only the fact that the analysis covered two alternative standards one slightly more stringent and the other slightly less stringent than the one EPA selected. It does not reflect the uncertainty associated with the estimation technique. Furthermore, there is no empirical evidence that children actually experience PDI, a symptom reported only by adults. See U.S. EPA (1996b, *55*).

25. See U.S. EPA (1997d, *38868*).

26. Ibid.

27. Ibid.

28. A careful reader may notice that the UV-B effects occur in the more distant future, because they result from a longer period of exposure. While this implies they are less important, EPA did not use such arguments in assessing the health improve-

ments from national ambient air quality standards (1996b). In addition, EPA did not use this argument in its most recent proposal (2001b).

29. Ibid. (2003, *634*).

30. See U.S. Court of Appeals for the District of Columbia Circuit (1999, *1053*).

31. See Tolley et al. (1994) for a summary and comparison.

32. See, e.g., Ames and Gold (1996); Hendee (1996).

33. See Shogren (1998a). U.S. EPA (1997f, *ES-9*) disagrees, writing, "[T]he $10,000 cost estimate for these reductions is intended to provide ample margin to account for unknown factors associated with future projects, and *may tend to overestimate* the final costs of attainment [emphasis added]."

34. The data are in an Excel file Case1i.xls available in the EPA docket. These data were unavailable when CEA staff prepared the $62 billion estimate.

35. This discussion is based on Lutter (1999b).

36. The data reject the hypothesis that it is zero with greater than 99 percent confidence.

37. It is not clear from EPA's analysis that these levels of emissions correspond to the least-cost way of meeting the standard. See U.S. EPA (1997f). In particular, reducing NO_x emissions a little bit more and VOC emissions a bit less may reduce the total cost of meeting the air quality standard. In the absence of information about such tradeoffs, I use EPA's estimates of the necessary emissions reductions. A recent scientific study indicates that this assumption is too optimistic; EPA's estimates of emissions reductions necessary to meet the standard in Los Angeles are in fact much too low. See Winner and Cass (2000).

38. See U.S. EPA (1997f, *9–15*).

39. Since EPA did not estimate the costs of attaining the standards for each metropolitan area, I had to derive EPA's cost estimates. I converted the reductions in tons of emissions per day during the ozone season necessary to attain the 8-hour standard, as presented in the Excel file Case1i.xls, to reductions in tons of emissions per year by multiplying by the number of ozone season days per year, given in the same file as 410 for NO_x and 329 for VOCs. I then multiplied by EPA's assumed cost per (annual) ton of $10,000 to arrive at the costs of reducing NO_x and VOCs to meet the standard in Los Angeles, $580 million and $340 million, respectively.

40. See Lutter (1999b).

41. This estimate reflects only retrospective estimates of cost declines.

42. See Winner and Cass (2000).

43. See Texas Natural Resource Conservation Commission (1999).

44. Total cost functions that are quadratic in (the log of) emissions reductions implied that annual costs for two cities exceeded $1 trillion per year; for another city, the cost exceeded $350 billion. The total cost for the remaining five cities I assessed was $19 billion per year. These estimates assume no technical progress. The regressions, however, have more coefficients that are statistically insignificant than the marginal cost functions presented here.

45. See U.S. EPA (1997f, *ES-17*).

46. The benefits of adopting all available control measures are $0.4 billion to $2.1 billion in 2010. Ibid.

47. Ibid. (9-6).

48. Grunwald (2000).

49. See Lutter (2003) and other testimony at that hearing.

50. See Browner (1999).

51. See Lutter (1999a).

52. See OMB (2002).

53. Ibid.

CHAPTER THREE

ECONOMIC ANALYSIS
AND THE FORMULATION
OF U.S. CLIMATE POLICY

Michael A. Toman

Over the past decade, the academic economics literature
on climate change has steadily grown (for reviews see Shogren and Toman
2000; Kolstad and Toman 2004). Economic analysis of climate policy also
is a significant part of the staff work in a number of government agencies.
The demand for economic analysis burgeoned as governments debated
policies to restrict greenhouse gas emissions (GHGs), both before and after
the December 1997 negotiation of the Kyoto Protocol (UNFCCC 1999b).

Economic ideas have played a noticeable and increasing role in formulating U.S. climate policy. That role, however, remained limited before and
after the Kyoto negotiations. Over time, economic analysts both within various government agencies and outside government have increased policymakers' confidence in the use of economic incentives as a powerful tool for
reducing GHGs. Economic analysis also has played an increasing role in
helping shape the international institutions that would be used to implement GHG controls, notably those institutions related to international

emissions trading. In sharp contrast, economic analysis has largely been ignored and occasionally even derided in the context of setting targets for GHG control. In particular, analyses of the costs and benefits of GHG control and of the most cost-effective timing of GHG limitation have been met with skepticism.

In my view, there are several reasons for this, each of which is important to bear in mind as one contemplates both the future of climate change policy and the broader role of economic analysis in the policy process. Some of these problems stem from the economic analyses themselves. For example, economic analyses frequently address only part of the policy problem.[1] Economic analysts sometimes have not been the most effective advocates for their own climate analyses.[2] One of the biggest obstacles to more effective use of economic analysis in climate policymaking, however, has been a basic lack of desire among many policymakers for the fruits of these analyses. This reluctance has been especially marked when the economic analysis clashed with strongly held preconceptions—on either side—about what climate policy ought to be.

CLIMATE ECONOMICS: SOME KEY CONCLUSIONS

1. There needs to be a balance of concern between the potential for irreversible negative consequences of climate change and the costs of misplaced mitigation investment.

Climate change is a real risk. But the size of the risk is unknown, and it will likely take several decades to become serious (IPCC 1998, 2001a, 2001b, 2001c). The uncertainty and time lag do not justify inaction. But they do underscore the need for a portfolio of actions to mitigate climate change risks, including adaptation as well as emissions mitigation efforts. Climate change is fundamentally a long-term problem, and long-term strategies are needed to address it.

Moreover, the extent to which the public will be committed to undertaking costly mitigation policies depends on the perceived risk. Uncertainty about the severity of the climate change risk will lower the intensity of public concern, especially relative to current policy matters seen as more pressing. The public's perceptions of the risk depend on how adversely climate change is seen to affect its descendants, and on how concerned the current public is for the well-being of its descendants versus its own well-being. It depends as well on how concerned people in richer countries are for the future welfare of people in poorer countries. These are complex issues of social values on which consensus is elusive (see Schelling 1995; Howarth 1998).

2. As part of a policy portfolio, a gradual but purposeful approach to the implementation of GHG control targets to take advantage of cost savings and opportunities for learning has many desirable features.

A gradual implementation of steadily more stringent GHG control targets causes less premature obsolescence of existing capital, and it provides more time for new, lower-cost mitigation methods to be developed and disseminated. It also allows society to learn more over time about both the risks of climate change and the options for responding. Critics of this view argue that without the necessary commitments to more substantial GHG reductions today, the problem just grows larger and the incentive to continue pushing it off into the future is increased. Moreover, they argue, aggressive abatement activity may help stimulate new technologies for reducing GHG emissions. In any event, current decision makers cannot force their successors to be good environmental stewards. And unless the risks of climate change are suddenly found to be very large or the public becomes very altruistic toward future generations, the high cost of excessively rapid action in the near term is a major deterrent to mitigation. It is better to simultaneously begin the work of mitigating GHGs while also making the job of more aggressive abatement easier for future decision makers through investments to develop technology for long-term GHG mitigation (for a review of these arguments, see Toman et al. 1999).

3. Well-designed, cost-effective GHG mitigation policies are essential. Thus, incentive-based mechanisms warrant a warm embrace, both domestically and internationally.

This economic proposition is widely accepted in the climate policy debate, at least in principle. It is now widely understood that no interest is served by needlessly cumbersome and inflexible policy tools, and that incentive-based tools such as carbon-based energy taxes and various emissions trading programs create powerful price signals for GHG abatement and technological innovation (Fischer 2001; Fischer et al. 2001; Pizer 2001; Wiener 1999a, 2001). There does continue to be strong debate over how broadly this proposition should be applied, including tradeoffs between economic cost-effectiveness and prerequisites for environmental monitoring. These debates are especially pronounced in the development of the "international flexibility mechanisms" for various kinds of emissions trading under the Kyoto Protocol (Hahn and Stavins 1999). In addition, there continue to be debates about the political feasibility of incentive-based policies, because they tend to make the price effects and overall costs of GHG mitigation more visible. But at the same time, these policies offer new options for addressing both distributional and efficiency concerns (Wiener 1999a; Goulder 2001).

4. *There are no doubt opportunities for devising and deploying improved technology, at relatively low cost, for GHG abatement. But technological solutions are not a panacea.*

Proponents of technology-based solutions argue that economic incentives are inadequate to change behavior to a degree sufficient to reduce climate risk. They advocate public education and demonstration programs; institutional reforms, such as changes in building codes and utility regulations; and technological mandates, such as fuel economy standards for automobiles and the use of renewable energy sources for power generation. They argue that the costs of such changes are negligible because the realized energy cost savings more than offset the initial investment costs.

This view, however, does not address several other factors that can cause the overall economic cost of a technology switch to be higher than an engineering-economic cost estimate based only on direct investment outlays and energy expenditures. For example, the new technology may be less reliable or incompatible with other existing energy-using processes. It is thus important to distinguish real market failures that impede low-cost choices, as opposed to market barriers that simply reflect unavoidable direct or hidden costs (Jaffe et al. 2001).

The role of government in lowering real market failures can include subsidies to basic research and development to compensate for an imperfect patent system; reform of energy sector regulation and reduction of perverse subsidies that encourage uneconomic energy use; and provision of information about new technological opportunities. Correcting market failures in the development of technology, along with creating appropriate market incentives for environmental improvement (point three, above), is expected to be superior to government selection of technology "winners." In developing countries, economywide policy distortions and infrastructure problems compound sectoral market failures, and solutions are found in broader institutional and market reforms, such as greater availability of information and expansion of financing opportunities, as well as reforms in energy sector pricing (Blackman 2001).

5. *Climate policies should be coupled to broader economic reform opportunities and other environmental policies to maximize win–win opportunities.*

When uneconomic subsidies that also encourage GHG emissions can be rooted out, this results in greater economic efficiency and better local environmental conditions, as well as lower GHGs. Similarly, when national trade and investment policies limit the spread of lower-GHG energy technologies or encourage uneconomic deforestation, there are clear potential opportunities for win–win reforms. Environmental policies to address national or local concerns also can reduce GHGs.

Some caveats are associated with these observations, however. Energy subsidies already have been substantially reduced over the past 10 to 15 years in both industrialized and developing countries; distortions remain, but the opportunity for a "free lunch" in reducing GHGs along these lines is diminishing. Moreover, many of the expenditures or tax breaks labeled as subsidies may not in fact contribute much to increased emissions, even if they add to energy producer profits—for example, tax breaks to small U.S. oil producers that do little to stimulate global consumption by lowering international oil prices (Fischer and Toman 2001).

With respect to broader trade, investment, energy, and industrial policies, no doubt opportunities for win–win improvement exist, especially in developing countries (Blackman 2001; López 2001; Toman et al. 2002; Toman 2003). The same is true of strengthened local environmental policies in developing countries (see, e.g., Morgenstern et al. 2002). But no one should underestimate the institutional difficulties and distributional consequences, and thus the political challenges, of such measures. It is also important that policy reforms make sense on their own. GHG abatement should not dominate the policy agenda.

6. *To address domestic political problems arising from the distributional impacts of GHG policy, efforts should be undertaken to compensate the greatest number of real losers with the least waste.*[3]

Losers from climate policy—those bearing costs that are disproportionate or in some other sense undue—can be compensated in a variety of ways. Domestically, if energy is taxed to reduce GHGs, businesses and households in those regions most harmed could get some offsetting tax relief. In a domestic tradable permit system, some permits can be given gratis as a form of compensation, or the government can auction the permits and redistribute some part of the proceeds, as with a carbon-based energy tax. To be cost-effective, such compensation schemes should be well targeted to limit compensation paid to those better able to protect themselves from the economic effects of GHG policies, and they should cause as little overall economic distortion as possible (Goulder 2001). With respect to targeting, care should be taken in evaluating the vulnerability of potential recipients of compensation. Low-income households or displaced workers may be more vulnerable than business managers operating in diverse markets or shareholders, who have even more opportunities for spreading their risks.

7. *A greater emphasis is needed on price-based approaches over strict quantity targets in the short to medium term to manage the risk of uncertain response costs.*

This key point links the design of GHG policies with the stringency or protectiveness of climate policy. The Kyoto Protocol established various fixed emissions targets for the Annex I industrialized countries. Meeting these targets no matter what technologies become available implies exposure to considerable uncertainty about the overall economic cost of the mitigation effort. If technological optimists are right about the potential for huge reductions at low cost, it will be easy to comply, but if new technologies are ineffective or expensive to deploy, then costs could accelerate rapidly beyond the initial cheap abatement opportunities.

An alternative approach is to set a domestic price ceiling for GHG abatement, a kind of safety valve that prevents such adverse cost outcomes (Pizer 2001). A carbon tax would automatically have this property: no one would pay more than the tax rate to abate emissions. A tradable permit system can be augmented by a safety valve by having the government offer additional permits at the specified ceiling price. Some practical complications must be addressed in implementing such policies, especially internationally, because international trade in permits makes sense only if participating countries all adopt the same safety valve.

A common critique of the safety valve approach is that it does not treat the quantitative targets agreed upon at Kyoto as immutable. But the sentiment for meeting the targets come what may does not seem that strong in practice. As illustrated by the Bush administration's decision to reject the Kyoto Protocol, and limited support within the Senate for ratifying it, there clearly is concern in the United States about high abatement costs. But it is also easy to envisage Europe and Japan and other Annex I countries making a best effort and then simply acknowledging failure if they do not succeed. After all, the Kyoto Protocol rests on national voluntary self-enforcement.

Loosening the emissions caps will do a bit less to slow climate change in the short term; given the long-term nature of the problem, abatement can be accelerated later if need be. A moderate slowdown in abatement is unlikely to have a major social impact, and if new information about climate change risks is obtained, the safety valve can be tightened.

8. Coherent international architecture is key to success. To this end, serious discussion is needed of common ground for common but differentiated participation of developed and developing countries based on shared burdens and mutual benefit.

The problem of achieving effective and lasting agreements can be stated simply: a self-enforcing deal is easiest to close when the stakes are small or when no other option exists (a clear and present risk). Climate change falls in between these extremes: nations have a common interest in responding to the risk, but each nation's incentive to reduce emissions is limited because it cannot be prevented from enjoying the fruits of other nations' efforts. Moreover, because no global police organization exists to enforce an international climate agreement, an agreement must be voluntary and self-enforcing—all parties must have an incentive not to deviate unilaterally from the terms—but this is hardest to achieve when incentives to free-ride are strong, as with climate change (Barrett 1994; for a discussion of alternative approaches, see Aldy et al. 2003b).

Equity is a central element in resolving this challenge. But no commonly held standard exists for establishing the equity of any particular allocation of GHG control responsibility (Cazorla and Toman 2001). Simple rules of thumb, such as allocating responsibility based on equal per capita rights to emit GHGs (advantageous to developing countries) and allocations that are positively correlated to past and current emissions (advantageous to developed countries), are unlikely to command broad political support internationally. Direct resource transfers from richer to poorer countries, or increased allocation of international GHG emissions quotas, likely are needed to increase developing countries' incentives to join an agreement (Wiener 1999a). But agreement on the magnitude of such transfers is tantamount to a more fundamental international agreement on what constitutes a fair distribution of responsibility for global GHG reduction. Achieving agreement on this issue will prove to be very difficult.

9. Adaptation measures need to be substantially strengthened, especially in developing countries.

Adaptation means increasing the resilience of natural and socioeconomic systems to respond to a changing and possibly more unpredictable climate system. Adaptation has until recently received relatively little diplomatic attention or financial support in the international climate policy process (though see Sperling 2003 for indication of growing attention to climate adaptation as part of poverty alleviation). Some environmentalists view it with some repugnance, seeing it as an overoptimistic excuse to avoid taking action to arrest climate change. I certainly do not believe that adaptation is a substitute for climate change mitigation. Nevertheless, adaptation is an important part of the policy portfolio. Some estimates suggest that adaptive responses by producers and consumers throughout the economic system can substantially reduce the effects of climate change, at least over the next century and where the necessary technical, institutional, and financial capacities are present (Mendelsohn 1999). These estimates remain somewhat controversial, but they highlight the importance of considering adaptation, especially because past and current GHG emissions already will lead to some degree of climate change in the future.

Adaptation potential is greatest when human involvement in the natural system is the greatest and when knowledge, infrastructure, and financial resources are the most plentiful. This means that adaptation is easier in agriculture than in wilderness protection, for example. It also means that adaptation is a greater challenge for poorer countries. Finally, a variety of policies can be implemented to strengthen adaptation capacity. A number of these policy measures are designed to overcome existing distortions and provide benefits in addition to reduced climate change risks. Examples include better water management, limiting encroachment in biodiversity-rich natural areas, and better zoning and building standards in coastal areas (Smith et al. 1996; Pielke 1998).

In summary, these nine points support an international climate agreement and domestic implementation program with the following properties:

- Moderate and flexible initial targets that accelerate over time.
- Flexibility in compliance strategies, with strong reliance on market-based mechanisms.
- Targeted and cost-effective compensation mechanisms for those whose cost burdens are judged to be excessive.
- Early and positive engagement with developing countries, both in exploiting win–win opportunities and in broadening commitments to global GHG control consistent with developing-country interests in economic progress.
- Strong support for research and development to create more cost-effective abatement strategies.
- Strengthening of opportunities for adaptation, as well as GHG mitigation, in industrialized and developing countries.

The Kyoto Protocol stands in sharp contrast to several of these points. While the protocol embraces in principle the use of incentive mechanisms, it also calls for substantial emissions control (for the United States, more than a one-third reduction in CO_2 relative to the expected business-as-usual emissions) to be achieved in a relatively short period of time to meet inflexible binding targets. The protocol also focuses primarily on near-term emissions abatement. It thus exacerbates the concerns of those who see a risk of excessive cost in meeting the protocol target, without a compensating benefit. Efforts to build support for safety valve proposals by economists thus far have had little practical success.

The compatibility of the protocol with the effective use of market mechanisms remains to be seen. The text of the protocol provides for the international use of various emissions trading mechanisms, both among Annex I countries and through emissions-reducing projects in developing countries that would generate tradable emissions credits—the so-called Clean Development Mechanism (CDM). The key modalities and procedures are still being debated, however, and strong sentiment exists among some participating countries to place significant procedural constraints on the mechanisms, out of fear that these mechanisms will unacceptably compromise the environmental integrity of the protocol. A large faction of countries seeks to substantially limit the overall use of the international mechanisms, preferring to see more domestic action that will stimulate longer-term emissions control. The goals of environmental integrity and longer-term GHG control are laudable, but those who advocate sharply limiting opportunities for cost-reducing emissions trading do not seem to fully appreciate the deleterious effects of this stance on environmental progress: by making participation in the shorter term more costly, they reduce incentives to participate in real emissions control.

The protocol also leaves to the future the issue of developing-country commitments to GHG limits. There is broad international acceptance of the principle of common but differentiated responsibilities for countries at various stages of economic development, but when and how developing countries would be engaged continues to be a source of dispute. In the current climate negotiations, debate focuses even more directly on the question of what financial and technical resources developed countries can and should provide to developing countries for GHG mitigation and adaptation. Moreover, provisions for effectively engaging developing countries in mutually beneficial GHG mitigation through the CDM remain undeveloped and therefore uncertain.

ECONOMICS AND U.S. CLIMATE POLICY

The issue of how international climate negotiations evolved is beyond the scope of this discussion. I will instead consider one piece of the puzzle: the way that economic analysis has—and has not—influenced climate policy in the United States. The analytical approach in this country to key climate policy questions has become steadily more sophisticated over the past 10 years. Nevertheless, the influence of this analysis on actual policy positions remains uncertain.

In the first Bush administration, the topic of climate change commanded interest at the highest levels. Nevertheless, some key figures in the administration were skeptical about the need for substantial reductions in GHG emissions because of the apparent scientific uncertainty surrounding the nature of climate change and the risks it might pose. That administration advocated actions that would have broad-ranging benefits, such as eliminating chlorofluorocarbons (CFCs) and other stratospheric ozone–depleting substances that were also GHGs (under the Montreal Protocol); implementing various pollution-control measures that also would promote energy efficiency (under the Clean Air Act); increasing biological sequestration of carbon in forest "sinks"; encouraging energy efficiency in buildings, appliances, and lighting; and increasing the use of renewable and nonfossil sources of energy. The broad-ranging benefits would include, in the judgment of some proponents, a decreased reliance on foreign oil imports and increased use of U.S. energy sources and technologies, as well as environmental benefits.

The United Nations Framework Convention on Climate Change (UNFCCC) was negotiated and ratified in 1992, during the first Bush administration (UNFCCC 1999a). That administration also drew attention to the economic dimensions of climate change and did a great deal of important stage-setting work for more cost-effective GHG mitigation policy. Among the issues addressed at that time were the need to manage multiple GHGs and sinks such as forests, not just energy-sector CO_2 emissions, so that net GHG abatement could be accomplished as cost-effectively as possible; and the importance of allowing for both international and domestic emissions trading policies, again for cost-effectiveness (Stewart and Wiener 1992).

In the wake of the successful negotiations to phase out CFCs and other substances that deplete the stratospheric ozone layer, some countries in the late 1980s began calling for significant reductions in global GHG emissions to be led by industrialized countries. But there was little momentum behind this advocacy at the time, and the UNFCCC contained only voluntary, nonbinding targets for industrialized countries to reduce GHGs to 1990 levels by 2000.

Although the Clinton administration came into office with stronger rhetoric about the need for GHG control policies, it began with voluntary technology-based measures promulgated in the 1993 Climate Change Action Plan (Clinton and Gore 1993). This approach was based on a premise that substantial progress toward reducing GHG emissions could be achieved without adverse economic consequences. The administration touted the economic benefits of cleaner and more climate-friendly technology. In practice, the voluntary technology-based approach reflected a compromise within the administration between those who believed there was a need for aggressive GHG control measures and those who were more skeptical.

The administration did, at various times in its first term, consider some modest legislative measures to help limit GHGs, such as policies to reduce tax benefits for employer-provided parking and efforts to provide support for mass transit. But a centerpiece of its early legislative efforts, the attempt to raise energy taxes (the "BTU tax"), went down in ignominious defeat in Congress. Only a small increase in the gasoline tax was achieved, and even this policy was hotly contested. Moreover, the energy tax measures ended up being sold to a large extent as budget policies, not environmental policies.

By 1996, it was clear that the United States would fail to achieve the voluntary UNFCCC goal of reducing emissions to 1990 levels by 2000. Reasons cited for the failure included less program funding than anticipated from a Republican congressional majority and, no less important, overoptimistic goals and a misspecified baseline (for example, oil prices did not rise as projected, and economic growth was unexpectedly rapid).

During the earlier years of the Clinton administration, several overlapping groups of policy proponents dominated the debate. One group of players in the development of administration positions tended to emphasize the risks that climate change could pose. A second group downplayed the importance of response costs by emphasizing potential "free lunch" opportunities in the technological arena. A third group emphasized the need for the United States to appear environmentally responsible in the international arena.

One consequence of this mindset was that there was virtually no interest at the senior policy level in the growing literature on the economic benefits and costs of GHG control (see Fankhauser et al. 1998 for a review of benefit and cost estimates). The great skepticism with which this literature was regarded within the political leadership of the administration was based on a view that the benefits of GHG control were being understated and the costs were being exaggerated.

One frustrating element for economists involved in the debate was widespread confusion between *total* and *incremental* benefits and costs. The standard economic models did not assert that GHG control should be avoided; they did imply that the presence of more low-cost and no-regret abatement opportunities argued in favor of more stringent control. The economic models cast doubt on the wisdom of many policy proposals circulating during the period, however, including stabilizing emissions at levels well below 1990 levels over a fairly short period of time.

Equally frustrating for economic analysts was the use by advocates of studies that purported to show that the total cost of reducing GHGs was low if a combination of energy efficiency standards and policies encouraging the voluntary adoption of lower-emissions technologies were pursued (see, e.g., IWG 1997). The low total cost of such policy packages was to a significant extent a consequence of a large "free lunch" from improved energy efficiency. Aside from doubts about the size and price of this lunch, economists rarely were successful in drawing attention to regulatory actions that had high incremental costs and thus were hard to justify, even if the total cost of the policy package was low.

Because of the mindset of administration policy leaders and the bad experience the administration had with proposals for energy tax increases, relatively little attention was paid to economywide analyses of GHG abatement costs or incentive-based policy instruments during the administration's early years. Discussions of carbon-based energy taxes and, to a somewhat lesser extent, emissions trading were to some degree taboo. Nor was there much interest among the political leadership of the administration in relating work within the agencies to economic research work being done outside of the administration on the effects and costs of climate policy, such as the work of the Stanford Energy Modeling Forum.

Instead, an elaborate bottom-up effort was undertaken to "find the tons" of potential CO_2 reductions by cataloguing specific technologically feasible abatement opportunities in various sectors. A large interagency working group led by White House offices was assigned the task of reviewing agency analyses of different abatement opportunities. The proposals, all offered in good faith and after hard work by the agencies, included improved building insulation or electric motor designs, and increased carbon sequestration on agricultural wastelands. The proposals were supposed to include estimates of the cost over time to achieve the carbon savings, but these estimates were spotty and involved only direct outlays for government programs, not the broader notion of economywide opportunity cost that is relevant for economic tradeoffs. In a number of cases, it was not even clear what policy levers would or could be manipulated to achieve the posited carbon savings, or how the proposed actions were additional to those already embodied in the 1993 Action Plan. Even efforts to create a kind of qualitative supply curve for the bottom-up GHG reductions by indicating the kinds of cost thresholds at which they might be possible were viewed as both too speculative (probably true, given the limited data) and too politically sensitive.

The administration's mindset—concern about environmental risk and diplomatic appearances, plus faith in massive win–win opportunities— along with the apparent failure of voluntary targets to accomplish what was desired led to U.S. support at the 1995 international climate meetings in Berlin for binding national emissions targets to be applied relatively quickly to developed-country emissions. The administration did not, however, spell out which goals or policies it would support, or what economic effects it expected to result from binding GHG emissions targets. During the two years of negotiations leading to specific emissions targets and a timetable for emissions control in the 1997 Kyoto Protocol, the administration was caught between domestic skeptics, many but not all Republicans, who questioned the need for the entire enterprise, and international diplomatic pressure for tougher action coupled with some pressure from domestic environmental groups. The Kyoto negotiations included a last-minute compromise by Vice President Gore to accept a stricter target. The U.S. delegation also pushed hard for the cost-reducing international flexibility mechanisms built into the protocol.

This last element of the negotiations indicated a growing appreciation of the importance of flexible, incentive-based policies at an international level. Economic analysis played a major role in building the case for these policies by showing their potential cost savings and by analyzing how the institutions governing the flexibility mechanisms might operate in practice (for example, how different liability systems for emissions trading would affect incentives to invest in and monitor the quality of abatement activity). Nevertheless, the Kyoto negotiations were inconclusive on the all-important details of how the mechanisms would operate, and subsequent negotiations have done little to close the gap. Moreover, there was an interesting schizophrenia in the U.S. policy debate: increased support in the U.S. policy process for international flexibility mechanisms did not automatically lead to increased interest in exploring domestic mechanisms. Domestic mechanisms that made the incremental cost of GHG control plainly observable were still seen as politically problematic.

In contrast to the discussion of means for GHG control, the process of setting targets for GHG emissions continued to be viewed mainly as a diplomatic, not an economic, matter. In particular, a striking and intellectually influential paper by Wigley et al. (1996) on the opportunities to achieve long-term GHG control more cost-effectively through more gradual and accelerating targets was largely discounted in the policy process as providing too much scope for abdicating responsibility. The U.S. negotiating position leading up to Kyoto did temporarily contain some limited provisions for borrowing against future emissions obligations as a way to smooth the path of GHG control, but this approach was roundly condemned by other countries and dropped in the final stages of negotiations.

The safety valve concept also was raised with the administration, but it was rejected in the weeks leading up to the final Kyoto negotiations. The United States did maintain a position that developing countries should be asked as part of the protocol to implement improved domestic policies and practices that would reduce GHGs while apparently being in their own self-interest. This position also was dropped in the final Kyoto negotiations, however, when developing countries strongly criticized it.

The willingness of countries to actually implement tough standards—as opposed to just agreeing to them—was debated, but the Kyoto Protocol did not contain any concrete compliance provisions. Compliance provisions in international environmental and other agreements are inherently limited, given the unwillingness of individual countries to sacrifice too much sovereign authority. The complete absence of compliance provisions in the Kyoto Protocol, however, meant that despite the primacy accorded to environmental integrity in the negotiations, the actual environmental contribution of the agreement remained open to question. The lack of compliance provisions also meant that there was little to dissuade countries wishing to appear green from posturing, adopting strong rhetoric about the need for strict measures without a strong conviction to take tough measures in practice.

As negotiations proceeded toward the Kyoto Protocol, the U.S. Senate passed by a vote of 95 to 0 a nonbinding resolution offered by Senators Robert Byrd and Chuck Hagel in the summer of 1997. The Byrd–Hagel resolution first stated that the United States should accept no climate agreement that did not demand comparable actions by all participants, developed and developing country alike. This provision was stimulated by concern about the competitive effects on the U.S. economy of a climate agreement. It has led to much contention since its passage, however, since it contradicted the focus in the 1995 Berlin Mandate on developed-country activity. It arguably also contradicted the language about "common but differentiated responsibilities" for GHG control in the U.S.-ratified UN Framework Convention, as resolution supporters sought to have developing countries undertake GHG abatement policies that could have imposed significant costs on those countries.

Another provision of the Byrd–Hagel resolution that has received less attention stated that the administration had to provide documentation and justification for the economic costs that legally binding GHG targets would entail. This part of the resolution was a concrete signal to the administration of congressional interest in the economic implications of a climate change agreement.

After the Kyoto Protocol was negotiated, President Clinton stated that the agreement would not be sent to the Senate for ratification until policymakers settled disputes about policies for flexibility in the means of compliance, costs of compliance, and "meaningful participation" by developing countries. Acrimonious debate erupted sporadically between the administration and Congress, as well as among various nongovernmental stakeholders, about budgetary priorities related to climate change programs and the consequences of climate policies for the U.S. economy.

As pressure mounted on the administration even before the Kyoto Protocol was negotiated to assess the price tag for GHG control, the administration reconstituted an interagency analytical team, this time chaired within the Commerce Department. Attention continued to be paid to bottom-up cost estimates for specific abatement opportunities, and economists involved in the process continued to play a role in critiquing many of these analyses for being overoptimistic. But now there was a new focus on findings from various economywide models. The administration team worked with a small number of selected models, including a long-term equilibrium model of the U.S. economy, an engineering-economic model of the energy sector, and a macroeconomic forecasting model with a detailed energy sector representation. Janet Yellen, chair of the president's Council of Economic Advisers (CEA) at the time, was given the task of speaking for the administration on the economic assessment. Although the administration did engage in efforts to estimate the costs to the U.S. economy of GHG limitations, it repeated much of its earlier rhetoric about the immediate win–win economic benefits of GHG reduction.

The dilemma for the administration both before and after the Kyoto negotiations was that it felt it needed to take a relatively strong stand on GHG control on environmental grounds, but given the hostility of the U.S. public and Congress to higher energy prices, it could not say publicly that GHG control would significantly increase energy costs. The administration maintained throughout the Kyoto negotiations and thereafter that the targets being negotiated would not be unduly burdensome.[4] It was widely believed at the time that the administration had its own ceiling on what would be a politically manageable increase in the price of fossil fuels, though such a ceiling was not publicly stated or documented and its existence is a matter of speculation.[5] Expressed in terms of carbon content, the ceiling was reputed to be on the order of $20 per ton of carbon. This translates into about 5 cents per gallon for gasoline, about the same size as the gasoline tax increase enacted in 1993 after the fight over the BTU tax.

March 1998 testimony by Yellen provided a carefully hedged statement that the United States could implement its Kyoto target while remaining within an acceptable range of energy price increases, and with correspondingly modest effects on the economy as a whole (Yellen 1998). In response to criticism that the administration did not provide documentation of its conclusions, the CEA released a July 1998 report (CEA 1998) stating that under the most favorable policy circumstances, the effects on energy prices and the costs to the United States of meeting the Kyoto Protocol emissions target could be extremely small. These costs were "likely to be modest if those reductions are undertaken in an efficient manner employing the [various international] flexibility measures for emissions trading." By modest, the administration report meant an annual gross domestic product (GDP) decrease of less than a few billion dollars per year, or about 0.1 percent of GDP; no expected negative effect on the trade deficit; increases in gasoline prices of only about 5 cents a gallon; lower electricity rates; and no "significant aggregate employment effect."[6]

Critics labeled the report unduly optimistic and out of step with mainstream economic analyses. Although the figures are theoretically possible, they are quite optimistic because they assume a very high degree of success in implementing the Kyoto Protocol flexibility mechanisms, especially emissions trading with developing countries and the former Soviet Union. The CEA analysis presumed extremely heavy use of emissions trading by the United States to comply with its Kyoto target, and an extremely efficient market in which this international trading would take place. Developing countries such as China and India were assumed to be very active participants in supplying emissions permits, even though they do not have emissions obligations under the protocol, and even though many developed and developing countries have expressed significant reservations about supplying large quantities of emissions permits to Annex I. Moreover, political constraints on the transfer by the United States of billions of dollars per year to developing countries and to Russia (which is likely to be a large permit supplier, as its Kyoto allocation exceeds its expected emissions in 2008) were ignored. And transactions were assumed to be costless, notwithstanding the likely procedural burdens of international transacting.

The CEA numbers also can be seen as optimistic in comparison with the results of an Energy Modeling Forum study of the cost of meeting the Kyoto targets (Weyant and Hill 1999). Consequently, the CEA report did not quiet the debate in the United States over the actual cost of meeting the Kyoto target, and whether the cost was worth it.

In fairness to the analysts at CEA and other agencies that worked on the 1998 report, it should be noted that the report does contain some information about the costs of achieving the Kyoto targets under less optimistic scenarios. Specifically, Table 4 on page 52 of the report includes information about the percentage reduction in GHG permit prices and direct compliance costs under different scenarios for international emissions trading, relative to a domestic-only approach. From this table and other information in the report, one can construct the cost of a domestic-only approach, which is several times that under optimistic trading assumptions. From this calculation, one can guess at the magnitude of costs under limited trading as a result of various institutional and policy frictions. But the fact that the information about domestic-only action is provided so circumspectly in the report speaks volumes about the political climate in which CEA and other analysts were operating.

In the wake of the Kyoto Protocol negotiations, economists in and out of government were ensnared in issues related to the design of GHG abatement policies as well as discussions of overall costs. Here economists in and out of government have been able to make some important contributions. For example, economic analysis has underscored the counterproductive character of "supplementarity" limits, quantitative limits on the use of the flexibility mechanisms: such limits would distort international economic activity by creating different carbon prices in each country, and they would threaten the entire operation of the mechanisms, as it would be difficult to determine whether any given transaction was in compliance with some relevant set of supplementarity constraints. Nor would they necessarily benefit developing countries, as they would limit demand for emissions credits through the CDM. But the last chapter in the story of international flexibility mechanisms has yet to be written. This is also true of the broader question regarding the influence of economic analysis on the outcome of the post-Kyoto negotiations.

The formal debate on the design of U.S. policies if the United States were to accept any GHG limits is even less advanced. There is some general acceptance of the idea that the United States would pursue a cap-and-trade system for limiting domestic CO_2, and perhaps to some extent other GHGs. But the coverage and structure of this system remain uncertain. Would permits be required only of electricity generating plants, and maybe other large fossil fuel users? Or would the permitting system attempt more comprehensive emissions coverage by applying "upstream," to fossil fuel suppliers?[7] Another key question is whether some or all permits would be auctioned—thereby creating a highly visible and politically uncomfortable price signal—or whether they would just be supplied gratis.

Another strategy that environmentalists would like to pursue is an increase in vehicle fuel economy standards. Economists generally oppose this policy on the grounds of inefficiency, as it would affect emissions intensity but would not limit and may even encourage vehicle usage (Portney et al. 2003). Continuing strong political opposition to this strategy exists as well. The Senate rejected an effort to mandate substantially more stringent vehicle efficiency standards in February 2002. The alternative of fuel taxes remains politically toxic, especially in the wake of the energy cost jumps in 1999–2001. More broadly, how much will the United States seek to orient other policies—in particular, those involving conventional pollutant control, electricity restructuring, and energy security—to influence future GHG growth? How much of a further effort will be made through specific technology promotion measures to engender hoped-for low-cost GHG reductions?

In February 2002, the Bush administration announced its plan for voluntary measures to achieve an 18 percent improvement in emissions intensity, emissions per unit of gross domestic product, over the next decade.[8] This plan and its context are discussed thoroughly in Chapter 1 of this book. As Pizer points out, from an economic perspective, the plan has a number of positive features. The setting of the target arguably goes beyond business as usual in terms of GHG emissions, and it is broadly within the range of potential emissions limitations that would be justified under standard cost–benefit analyses of GHG mitigation targets. Moreover, the framing of the target in intensity terms avoids the drawbacks of rigid quantitative targets already noted above in the context of a safety valve approach; and the Bush plan would begin to put in place important parts of a policy infrastructure for more stringent policies in the future (such as GHG reporting capacities and multiple-gas approaches to take advantage of cost-effective reductions in GHGs besides CO_2, such as methane).

As Pizer also notes, however, the plan was not accompanied by any binding policy measures and consequences for nonattainment. The measures envisaged for achieving the emissions reductions come down mostly to limited tax benefits and industry "challenges" to take voluntary measures. These are the same sorts of measures that most economists found ineffective at achieving even limited reductions in the 1990s. Given the political aversion in both U.S. political parties to policies causing even modest increases in energy prices, this outcome is not surprising. Nevertheless, for economists who greeted the 1993 Clinton–Gore action plan with skepticism, the renewed prominence of voluntary actions is an irony.

CONCLUSIONS

Against this backdrop, what can be deduced about the capacity of economic analysis to affect U.S. climate policy? Generally, economists have been perceived as not having enough good news to offer. Economic analysis has offered a number of useful ideas for increasing the cost-effectiveness of GHG policies, but it also has emphasized the near-term costs of GHG control, as well as long-term benefits, and the importance of adaptation to climate change as a risk-mitigation measure. This emphasis has been seen by many advocates in the debate as incorrect, both factually and philosophically.

Under these circumstances, it is difficult to develop a constituency for good economic analysis within the policy process. Green politics—pressure from domestic environmental constituencies and international diplomats—overwhelmed the counsel of economists during the Clinton administration. In a similar vein, brown politics—the desire to oppose any controls so as to avoid a debate about what level of stringency is desirable—appear to have prevented the current Bush administration from adopting modest mandatory measures. Certainly the energy policies of this administration are open to criticism if it takes no real action at all to create a price incentive for reducing GHGs and instead peppers the tax code with costly special benefits to subsidize specific new technologies. Economists in government agencies such as CEA cannot be, as one colleague put it, skunks at the picnic, or else they will not get invited as they would like to meetings on other important issues.

Given the substantive uncertainties and political controversies surrounding complex environmental issues such as climate change, it is important to preserve an independent, not overly politicized capacity for economic analysis within the executive branch. CEA is unique within the U.S. executive branch in terms of being able to provide high-quality and objective, yet timely and relevant, economic analysis of such policy issues.[9] A number of other agencies, including the Environmental Protection Agency (EPA), the Energy Department, and the Treasury Department, also play important roles in this process. It is hard to see what virtue is served over the long term by limiting the flow of good economic information and analysis or by channeling analysis only to satisfy immediate political ends.[10]

From these observations, I draw two sets of recommendations for strengthening the quality and credibility of administration economic analysis of climate change. First, there is a need for continued close links between government policy analysts as consumers of economic analysis and the world of nongovernmental research. This link has become stronger over time, in part because of government financial support for climate change economics research. CEA also can play an important role in maintaining these links, given its own ties to the research world and its ability to impartially referee analytical disputes. To play this role, however, CEA needs the organizational and financial capacity to convene workshops and commission reviews of the state of knowledge.

The other set of recommendations concerns internal safeguards against abusing or ignoring economic analyses in evaluating climate policies. In 1996, the Office of Management and Budget (OMB) published updated guidance for agencies to use in carrying out their own economic assessments of proposed rules. In 2000, OMB again issued updated guidance, and in 2001, the new administration admonished agencies to adhere to the guidances issued by the earlier administration (see Lew 2000; Daniels 2001). One can debate how well this guidance has been followed by agencies in practice, but the ideas expressed in it nonetheless carry considerable intellectual weight. Among the provisions in the guidance are the following points:

- Regulators should seek to quantify risks, benefits, and costs and monetize these values to the greatest extent practicable.
- Regulators should consider realistic baseline conditions and make realistic evaluations about the effects of proposed policies.
- Regulators should consider a range of plausible goals and actions.

What if these general principles were truly applied not just to agency rules, but also to administration policy pronouncements? At a minimum, the 1998 CEA report would have been different, and the Bush administration's recent decision to adopt purely voluntary measures might have been accompanied by an assessment of the increased CO_2 concentrations that will ensue from the delay in controlling emissions. More to the point, policy statements on climate change would be backed up with a degree of analysis that would add to their credibility, and filter out those that are not really that credible. CEA could play a critical role in this process by testing the intellectual integrity of different ideas, bringing in relevant outside analysis, and providing or coordinating background material in support of those assertions that have successfully passed through this relatively coarse filter. An approach like this to climate policy might enhance what all sides in the debate advocate: reliance on good science.

NOTES

I am very grateful to John Anderson, Howard Gruenspecht, Randall Lutter, Adele Morris, Jason Shogren, and Jonathan Wiener for comments on previous drafts of this paper. I am entirely responsible for its content, however.

1. The focus of many long-term analyses on the present value of expected net benefits from policy does not address winners and losers across space or time, and analyses sometimes idealize policy institutions and thereby overstate the benefits of more incentive-based approaches. But these are not inherent weaknesses of the economic approach; they can be rectified by broadening the scope of analysis.

2. In particular, economists have not always made the point that *any* climate policy decision involves some tradeoffs, explicit or implicit. The debate has instead been framed by those who view either environmental protection or compliance cost avoidance as an objective beyond compromise.

3. Distributional issues also arise internationally; see point eight.

4. A White House press briefing after the protocol signing in Kyoto included remarks by Treasury Secretary Lawrence Summers that the protocol would be good for the economy, as well as laudatory statements by environmental officials in the administration.

5. The irony of this possible political safety valve was not lost on those economists who had advocated for an explicit and transparent ceiling price on GHG abatement costs.

6. The pre–Kyoto Protocol results from the interagency analysis team also are within this range (see Yellen 1998). One exception is that the earlier estimates involving a reduction of emissions to 1990 levels by 2010 would cost Americans 900,000 jobs by 2005 and 400,000 jobs by 2010. In the July 1998 report, a more ambitious emissions control objective was found to have a lesser effect on employment.

7. CO_2 from the power sector is only about one-third of the total U.S. emissions of CO_2.

8. A description of the administration's proposal can be found at http://www.whitehouse.gov/news/releases/2002/02/20020214.html (accessed June 3, 2004).

9. CEA is a small agency with no permanent staff and limited resources. Its sole institutional agenda is to bring economic insight and analytical rigor to administration policy debates.

10. In particular, if the reputation of CEA for objectivity is compromised, it will not be easy to rebuild, and there will be a larger cost to an administration in terms of policy credibility.

CHAPTER FOUR

SAVING THE PLANET COST-EFFECTIVELY: THE ROLE OF ECONOMIC ANALYSIS IN CLIMATE CHANGE MITIGATION POLICY

Joseph E. Aldy

No environmental problem likely poses a more significant policy challenge to decision makers than global climate change. The characteristics of the global climate change problem illustrate the difficulty in developing policy. The earth's atmosphere is a global public good: any successful effort to reduce the flow of greenhouse gas emissions to the atmosphere will likely require a coordinated international effort. Firms and individuals will incur the costs of mitigating climate change in the near term, but the benefits will accrue to future generations. Substantial uncertainty affects estimates of benefits and costs of mitigating climate change and also hampers our understanding of the effectiveness of various policy instruments. The risks of climate change have inspired many in the environmental community to focus their energy on this issue, while the costs of mitigating climate change have likewise motivated many in industry who fear future regulatory costs.

The significant complexities of global climate change reveal the need for informed policy deliberation and development. Decision makers can benefit from a better understanding of scientific, economic, technological, diplomatic, and political factors related to global climate change. Economic analysis can inform decision makers on a variety of important issues, including costs and benefits, characterization of uncertainty, and distributional effects. Such analysis can inform decisions on both the setting of policy goals and the means of achieving these goals.

The Clinton administration decided to present its own economic analysis of the Kyoto Protocol in congressional testimony by Janet Yellen, chair of the Council of Economic Advisers (CEA), three months after the Kyoto negotiations. The administration elucidated the results of the analysis underlying Yellen's testimony in a July 1998 report titled *The Kyoto Protocol and the President's Policies to Address Climate Change: Administration Economic Analysis* (hereafter referred to as AEA). The AEA bore significant criticism on two important assumptions: international emissions trading and developing-country participation. These assumptions, however, did not relate to economics, but to the outcomes of subsequent negotiations.

These criticisms are consistent with broad concerns about the limited practical influence of economics in practice on global climate change policy.[1] For example, several economists have criticized the Kyoto Protocol for failing a benefit–cost test (Nordhaus and Boyer 1999; Nordhaus 2001), for imposing greater costs than necessary for achieving possible long-term concentration targets (Manne and Richels 1999), for creating inadequate incentives for participation and compliance (Barrett 1998), and for forgoing cost-effective emissions reductions in developing countries (Bernstein et al. 1999). Achieving an agreement that may suffer from such economic flaws may be all the more surprising given the significant amount of economic research undertaken in academia, the private sector, nonprofit organizations, and government agencies prior to the Kyoto negotiations (e.g., Gaskins and Weyant 1993; Novak 1997; Alliance to Save Energy et al. 1997; and IAT 1997).

The administration's economic analysis of the Kyoto Protocol reflected the history leading up to the Kyoto negotiations, including domestic policy development, international negotiations, and preceding economic analyses. More interesting, subsequent policy efforts reflected the administration's economic analysis, or, to be more exact, the cost-effective policy objectives underlying the analysis. The economic analysis raised the cost of changing administration policy and constrained those who suggested that the administration take a path of less resistance, which usually implied unnecessary costs or "doing it dumb." This chapter assesses the effects of the administration's economic analysis to illustrate the role that a published economic analysis can have on subsequent decision making.

THE ROAD TO THE KYOTO NEGOTIATIONS

"A great, great deal has been said about the weather, but very little has ever been done."

—Mark Twain[2]

The road to Kyoto began with the 1992 Earth Summit in Rio de Janeiro, Brazil, where about 150 countries signed the Framework Convention on Climate Change (FCCC). Among other provisions, the treaty specified non-binding emissions goals for the end of that decade for 40 industrialized countries[3] (Article 4) and an ultimate objective of stabilizing greenhouse gas concentrations at a "level that would prevent dangerous anthropogenic interference with the climate system" (Article 2). The FCCC made an important distinction between industrialized and developing countries, noting their "common but differentiated responsibilities and respective capabilities" (Article 3). Within the FCCC, this differentiation translated into nonbinding goals for industrialized countries and no goals for developing countries, developed-country financing of developing-country investments in climate-friendly technology, and more extensive and frequent reporting requirements for industrialized countries. Ratified with the advice and consent of the U.S. Senate in October 1992, the FCCC entered into force in 1994. By February 2004, 187 countries had become parties to the FCCC.

At the first conference of the parties to the FCCC in 1995, the world came to agreement on the so-called Berlin Mandate.[4] This mandate provided for a new round of negotiations for emissions commitments beyond 2000. Consistent with the precedent in the original treaty for "differentiated responsibilities," this process focused on commitments exclusively for the industrialized world. With the aim of achieving an agreement on a new set of emissions commitments by the third conference of the parties in 1997, the parties initiated an intense series of negotiations. At the 1996 conference in Geneva, the Clinton administration announced its willingness to accept binding emissions commitments under the Berlin Mandate process. Diplomatic efforts focused on the path toward binding emissions quotas for industrialized countries.

As the negotiations proceeded in 1996, the administration initiated an interagency process to assess the economic effects of emissions abatement policies. Dubbed the Interagency Analytical Team (IAT), this group of government analysts used four different models to evaluate the economic effects of various climate change policies, with a focus on the United States (see IAT 1997).[5] Whereas the academic literature at that time included a variety of analyses to determine the optimal climate change policy (e.g., Cline 1992; Nordhaus 1994) and the least-cost emissions path to a concentration stabilization goal (e.g., Wigley et al. 1996; Weyant 1997), the IAT focused on evaluating the costs of specific emissions targets for 2010.[6] The IAT presented results in terms of the price for a greenhouse gas tradable permit, energy prices, gross domestic product (GDP) loss, and unemployment.

In spring 1997, the administration compiled a "blue ribbon" panel of economists from academia, the private sector, and nongovernmental organizations (NGOs) to review the work of the IAT. A majority of the panel was very critical of the IAT's draft report, including its focus on a limited number of policy scenarios, its optimistic technology assumptions,[7] its lack of a benefit–cost analysis, and particular weaknesses of the models supporting the report.[8] In light of the critical peer review and growing pressure from Congress for the administration's economic analysis, the Clinton administration decided to release the IAT's draft report at a July 1997 hearing before the House of Representatives' Energy and Power Subcommittee (1997). In the first of her 10 appearances at congressional hearings on climate change policy, Janet Yellen testified on the IAT report and the administration's approach to economic analysis. Yellen discussed the lessons learned from the IAT and cautioned that attention should focus on the common results from the breadth of the modeling literature and avoid concentrating on the specific outputs of any one model.

One aspect of the IAT work merits additional comment. The IAT evaluated tradable emissions permit programs only. This reflected a strong interest within the administration for market-oriented implementation policies. The focus on tradable permits as the primary mechanism for achieving emissions quotas in lieu of technology or performance standards under a more traditional command-and-control regulatory approach represented a significant departure from previous environmental policy debates. Cost-effective implementation became a central principle of the administration's climate change agenda in both domestic and international arenas.[9] In terms of international policy, the IAT effort complemented previous research (Manne and Richels 1992; Bruce et al. 1996; Nordhaus and Yang 1996) that showed the potential for substantial cost savings through international emissions trading, both within Annex I countries and in a global system. These findings strengthened the hand of the advocates for international emissions trading, an idea that faced very little dissent within the administration.

From a political perspective, the IAT report was considered an incomplete and unsuccessful attempt to quantify the economic effects of U.S. climate change policies. The abandonment of more than a year of work on the report affected the policy development process in two ways. First, it frustrated members of Congress who were waiting for the administration's economic analysis. Second, it created a large hole: the Clinton administration did not have a public analysis to inform the debate over the negotiating position it would take to the Kyoto conference.

On this first point, the Senate responded soon after the demise of the IAT with a 95 to 0 vote for the Byrd–Hagel resolution (Senate Resolution 98). Although it was not binding, this resolution expressed the Senate's view that it would not provide its advice and consent to any agreement that did not meet the following two conditions: (1) developing countries must agree to emissions commitments in the same compliance period as industrialized countries and (2) the agreement must not cause serious harm to the U.S. economy. Further, the resolution requested that a report on the costs of implementing the climate change agreement accompany the protocol when submitted to the Senate for its advice and consent to ratification. Both the second condition and the request for an economic analysis demonstrated the frustration Congress had with the administration regarding the failure to present a climate policy analysis. Moreover, the Senate had been able to frame the policy debate in terms of developing-country participation and the economy. The domestic politics, focused on concurrent participation by developing countries, strikingly conflicted with the ongoing international negotiations under the Berlin Mandate.

On the second point, halting the IAT process created a void, and nature abhors a vacuum. Several of the Department of Energy's national labs collaborated to produce a report on possible emissions scenarios that essentially dressed up a technology inventory and some wishful thinking in economic jargon.[10] The so-called "5 Labs Study" made the claim that with a "vigorous national commitment" the United States could reduce its carbon dioxide emissions to its 1990 level by the year 2010 with no net cost to the economy (IWG 1997). Although this analysis lacked any representation of economic behavior, the macroeconomy, or for that matter, any identification or description of the policies that might constitute a "vigorous national commitment," it provided ammunition for advocates of ambitious climate change policies.[11]

In October 1997, President Clinton announced the administration's climate change policies. These included the administration's positions going into the Kyoto negotiations—1990 emissions target for the 2008–2012 commitment period, international emissions trading, and "meaningful participation by key developing countries"—as well as a set of domestic policies to promote emissions abatement and sequestration. The target certainly reflected the influence of the optimistic technologists. Including international emissions trading and pressing for more participation by developing countries, however, reflected a concern for cost-effectiveness. The domestic policies also reflected a split: in the near term, largely voluntary activities represented the technologists' view; for the 2008–2012 period, the proposal focused on a domestic emissions trading program. Given the world's response to this proposal in pre-Kyoto negotiations in November 1997, and the significant distance between negotiating blocs on virtually every issue on the table, the prospects for an agreement at Kyoto appeared dim.

THE ADMINISTRATION'S ECONOMIC ANALYSIS OF THE KYOTO PROTOCOL

> "I am particularly pleased the agreement strongly reflects the commitment of the United States to use the tools of the free market to tackle this difficult problem. There are still hard challenges ahead, particularly in the area of involvement by developing nations. It is essential that these nations participate in a meaningful way if we are to truly tackle this global environmental challenge. But the industrialized nations have come together, taken a strong step, and that is real progress."
> —President William J. Clinton, commenting on the Kyoto Protocol[12]

In December 1997, the Clinton administration had a hard-fought international agreement to address climate change but no public analysis about its effects on the U.S. economy. The economic team had undertaken a number of analyses "on the fly" during the last week of the negotiations as new proposals were tabled, but the final agreement had not been fully evaluated, in large part because several of the provisions in the protocol were not immediately amenable to the energy-economic models (e.g., inclusion of carbon sinks and non–carbon dioxide greenhouse gases). The administration quickly recognized that an informed public debate over the Kyoto Protocol could not proceed without an economic analysis of the agreement. Even if the administration did not present its own economic analysis, advocates on both sides of the issue would likely do so (and they did: see Novak 1998; DRI 1998; and Geller et al. 1999). The administration, therefore, decided to fill the void and undertake its own economic evaluation of the Kyoto Protocol.

The decision to conduct this economic analysis did not reflect any U.S. treaty law obligation, legislative mandate, or executive branch policy. The administration had the discretion to release a public analysis of the economic effects of the Kyoto Protocol. After the Kyoto negotiations, senior policymakers discussed how the administration would approach the ensuing debate on the agreement. It was decided that an economic analysis would play a central role in the administration's efforts to inform Congress and the public about the Kyoto Protocol. Although the administration could have opted against a public analysis, and possibly referenced the 5 Labs Study as justification for the Kyoto Protocol, it made the right decision to go forward with the AEA.

The administration's economic analysis included contributions by a number of agencies, although CEA took the lead and Chair Yellen served as the administration's spokesperson on the economics of the Kyoto Protocol. Yellen presented the results of the administration's economic analysis at a March 1998 congressional hearing.[13] The written testimony described the economically relevant elements in the Kyoto Protocol, reviewed the literature on the benefits of mitigating climate change, surveyed many of the key findings in the economics literature (such as the cost savings from trading), and provided an illustration of potential economic effects from a global economic model. Not surprisingly, attention focused on the numbers provided in two "illustrative scenarios": $14 to $23 per ton of carbon equivalent for a tradable permit in 2010. To complement the March testimony, in July 1998 the administration released the AEA. This analysis elicited criticism for being overly optimistic about the costs of implementing Kyoto.[14] The analysis arrived at a tradable permit price of $23 per ton, depending on several critical assumptions; these merit detailed discussion, especially in light of how they influenced subsequent policymaking.[15] The two key assumptions driving the analysis were efficient international emissions trading and "meaningful participation by key developing countries."

The administration constructed its estimated permit prices from a modified version of the second generation model (SGM).[16] The SGM, like virtually every global energy-economic model, solves for a carbon price by seeking the lowest-cost emissions abatement opportunities in all countries with emissions commitments. Effectively, it assumes that all countries with emissions commitments implement frictionless domestic tradable permit systems integrated with a frictionless international emissions trading system. All countries face the same price for carbon, and compliance with emissions commitments occurs at least cost.

How plausible is this frictionless trading assumption? The protocol provided limited guidance on this issue, as the international trading article consisted of three sentences. Hahn and Stavins (1999), Wiener (1999a), and others discuss the pitfalls in designing international emissions trading systems, and how a myriad of domestic programs could evolve over time that would complicate the attainment of global least-cost emissions abatement. It is not, however, immediately obvious how to incorporate inefficiencies in program design into a top-down global energy-economic model. Although virtually all economists who have developed global energy-economic models note that their tools assume frictionless trading, none had at that time (or subsequent to then) analyzed an international emissions trading system with frictions. For example, the Stanford Energy Modeling Forum's evaluation of the Kyoto Protocol published in a special issue of the *Energy Journal* in 1999 (MacCracken et al.) included 13 global energy-economic models; none of them modified their trading assumptions to account for potential inefficiencies in trading.[17]

Creating an asset worth hundreds of billions of dollars annually and then developing the rules for international trade in this asset over a decade is a daunting, formidable task. The world has spent more than half a century on rules for trade in goods and services, and the work still continues. Getting it right the first time around on international emissions trading will be difficult. Absent more detailed models of domestic economies and a better understanding of the emissions abatement programs countries will implement, however, the best modelers are likely to do is to bound their estimates with autarky on the high end and frictionless international trading on the low end.[18]

Given that the analysis assumes efficient international emissions trading, what countries are assumed to participate in trading? What does the phrase "meaningful participation by key developing countries" mean? In the case of the economic modeling, the analysis assumed that China, India,

Korea, and Mexico adopted emissions targets equal to their 2010 business-as-usual emissions levels (see the Subcommittee on Energy and Power 1998b, 296). The administration interpreted a first-period emissions commitment as sufficient, but not necessary, for meaningful participation. Although the administration never compiled a list of key developing countries, many recognized these four as very important. (As a former colleague put it, any country that takes on an emissions commitment is a key developing country.) The model also facilitated the analysis of these countries: the SGM has five developing-country modules—China, India, Korea, Mexico, and the rest of the world. Including these four countries, especially China, significantly reduces the costs of complying with the Kyoto targets for the United States and other developed countries. Coupling efficient international emissions trading with developing-country commitments would reduce the marginal cost of compliance with the Kyoto Protocol by nearly 90 percent in our analysis. Subsequent research by the Energy Modeling Forum (EMF) showed comparable gains. The 12 EMF models evaluating the U.S. emissions commitment under no international emissions trading and full global trading found on average an 80 percent decrease in marginal cost, while several showed cost savings in excess of 90 percent (Weyant and Hill 1999).

The assumption that these four developing countries would adopt emissions commitments contrasted with the reality of the Kyoto negotiations. The developing-country negotiating bloc opposed a mechanism to allow countries to voluntarily adopt emissions commitments in the negotiations. These countries did not have commitments and did not express an interest in adopting commitments, nor did the protocol provide them with a way to take on a commitment. Shogren (1998b, 11) notes that this "broad and deep" assumption about participation explains the "modest" cost estimates in the AEA. He questions this assumption, stating that "these estimates might be plausible if all goes exactly right with the world; a big 'if.'" Nonetheless, this developing-country participation assumption did conform broadly with the administration policy seeking more ambitious efforts by developing countries.

Some critics of the administration's analysis have claimed that it relied on an optimistic assumption about technological progress. The IAT incorporated an optimistic energy efficiency improvement assumption. The SGM, like many similar models, operates with an autonomous energy efficiency improvement rate (i.e., a rate chosen by the modeler). Whereas the IAT modeling runs reflected a choice of an optimistic rate of technological change, the AEA reflected a standard assumption made by many modelers. The analysis presented in Yellen's testimonies and the July 1998 report employs the baseline energy efficiency improvement value (about 0.9 percent per year improvement) used by the Energy Information Administration (EIA), which is broadly accepted in the modeling community.[19]

Critics claim that the economic analysis was overly optimistic and that the assumptions of efficient international emissions trading and developing-country participation were different from the likely outcome of negotiations. The analysis showed, however, that attainment of specific diplomatic policy goals was necessary to ensure that the Kyoto agreement would impose only "modest costs" on the U.S. economy. It raised the bar for the ratification package that the administration presumably would submit to the Senate at some point in the future.[20]

In contrast to private analysis for decision makers, in which economic analysis shows primarily economic tradeoffs, public analysis reveals economic and political tradeoffs. These tradeoffs in public analysis may not be so transparent to the casual observer, but they exist and are substantial. With a public economic analysis, changing a policy position requires more than fancy rhetoric, repackaging and rephrasing bullet points, and spinning the new policy. The new policy position may conflict with the existing analysis and require either a new economic analysis or an explanation of the policy change's effects on the conclusions of the existing economic analysis. This raises the political cost of changing the course of policy and creates a bias toward maintaining the cost-effective policies underlying the economic analysis. For example, consider a hypothetical case in which the administration had submitted the Kyoto Protocol to the Senate for its advice and consent to ratification and claimed that a joint statement with China about energy efficiency and renewable energy constituted meaningful participation.[21] The Senate would (should) force the administration to (1) defend the claim that this action is meaningful and (2) describe how it affects the cost estimates in its public economic analysis. Absent the economic analysis, the administration would have only to fight a battle of words over the definition of "meaningful."

The AEA may not have represented a textbook economic analysis such as those students read about in a public policy class. For example, the report did not provide a full benefit–cost analysis. Although some would argue that such an analysis should be undertaken,[22] several significant uncertainties would make drawing conclusions from such an assessment very difficult. First, a benefit–cost analysis would require an assessment of the emissions reductions from policies over the next hundred-plus years. The international community had specified emissions commitments only through 2012. Any assessment would then require arbitrary assumptions about future emissions paths for both Annex I and non–Annex I countries. Second, such an analysis would require the discounting of future benefits and costs. Given the very long time horizon for the benefits, the choice of a discount rate is critical. Recent research illustrates that such uncertainty should influence the choice of discount rate and significantly affect the present value of long-term benefits (Newell and Pizer 2001, 2003; Weitzman 2001).

In addition, some observers noted that the report did not provide an evaluation of alternative policies. Although assessments of domestic legislation and regulation certainly benefit by a consideration of alternatives, the value of such a comparison of options may be less for an international treaty. In some sense, a negotiated multilateral agreement offers only two options: take it or leave it. Such an evaluation of alternatives certainly would have been valuable *before* the Kyoto negotiations, but unfortunately, the ill-fated IAT addressed only three (very similar) policy options. Certainly, domestic climate change mitigation policies should be subject to an evaluation of alternatives. The Clinton administration, however, did not prepare an evaluation of alternatives when proposing several near-term policies to address climate change, such as the Climate Change Technology Initiative.

Despite these warts, the AEA did inform the public about the key elements underlying "modest cost" attainment of the Kyoto commitments. Although critics may have wanted more from an analysis, the Clinton administration provided more information on the economics of the protocol than most other governments. Whereas the Clinton administration chose to publish an economic analysis, the European and Japanese governments did not present any public economic analyses of the Kyoto Protocol over the 1998–2000 period. Having conducted and published the AEA, policymakers had some ammunition to advocate cost-effective policies on the domestic front, with the EU on international emissions trading, and with developing countries on their potential participation.

DEVELOPING DOMESTIC CLIMATE CHANGE POLICIES

"[C]osts depend critically on how emission reduction policies are implemented, and it boils down to this: If we do it dumb, it could cost a lot, but if

we do it smart, it will cost much less, and indeed it could produce net benefits in the long run."
    —CEA chair Janet Yellen, testifying at a 1997 congressional hearing[23]

How does one define "doing it smart"? In the case of the Kyoto Protocol,
economists would likely envision a "smart" policy as one that facilitated
cost-effective attainment of the Kyoto target. Any evaluation of a proposed
emissions abatement or sequestration policy would necessarily involve a
comparison with other alternatives to determine whether the proposal
passes a cost-effectiveness test. By publicly stating that the Kyoto Protocol
could cost $23 per ton, and reinforcing it with claims of "modest" costs,
the administration had effectively established the standard by which to
judge domestic policy proposals. "Could" easily became "would" as interpreted inside the Beltway, so any policy that would result in a marginal cost
of abatement less than $23 per ton would be cost-effective. Any policy
exceeding that value would not be cost-effective. Or put another way, policies with marginal costs below $23 per ton would be "smart," and those
exceeding $23 per ton would be "dumb."

This simple characterization of cost-effectiveness potentially influenced
domestic policy through both external pressure and internal deliberation.
The administration's proposed policies on tax credits and research and
development (R&D) funding for renewable energy and energy efficiency,
referred to as the Climate Change Technology Initiative (CCTI), failed to
garner congressional support in part because of questions about cost-effectiveness. In 1997, the president had announced the CCTI as a way to
"prime the pump" for technological development and deployment, and
this served as the core of the first part of his three-stage plan to address climate change.[24] Elements of the CCTI package included tax credits for electric and fuel cell vehicles, rooftop solar photovoltaic systems, and wind-
based electricity generation, as well as R&D funding for combined heat and
power systems, renewable electricity sources, and energy-efficient housing.
The president originally proposed $5 billion over five years for the pro-
gram, although the administration requested more than $6 billion for
these tax credits and R&D for fiscal year 1999.

The CCTI drew significant criticism from Congress. Some of the opposition reflected the antipathy some members of Congress felt toward anything
potentially related to the implementation of the Kyoto Protocol; other
members raised legitimate questions about how much "bang for the buck"
the CCTI would deliver.[25] To determine the cost-effectiveness of the CCTI
proposals, Congressmen James Sensenbrenner and David McIntosh

requested that EIA conduct analyses of the administration's FY2000 and FY2001 CCTI proposals.[26] The results of the EIA (2000a) analysis make for imperfect comparisons with the administration's analysis of the Kyoto Protocol, as EIA estimated the average revenue reduction per ton of CO_2 abated, whereas the administration estimated the marginal cost of the last ton reduced to comply with the Kyoto target. Nevertheless, the EIA analysis can provide some sense of the relative cost-effectiveness of the CCTI proposals.

Three aspects of the EIA analysis illustrate how the CCTI would not promote cost-effective emissions abatement. First, the average revenue reduction per ton abated exceeds the Kyoto Protocol economic analysis permit price of $23 per ton for every tax credit evaluated. None of the proposed tax credits cost less than double the $23 value per ton, and some cost 15 to 20 times more per ton.[27] Second, the average revenue reduction per ton varies by a factor of 10 across various proposed tax credits.[28] If the variation in revenue reduction for the marginal ton abated is comparable to the variation in this average, this implies that some of the proposals should receive more funding (the low cost per ton tax credits) and others should receive less (the high cost per ton tax credits). Shifting the funding around could increase the amount of emissions abatement for the same overall funding package, thereby improving cost-effectiveness. Third, EIA estimates that for a majority of the proposed tax credits, more than half of the tax credit recipients would have invested in the targeted technology even without the tax credit. For example, 80 percent or more of tax credits for wind energy, rooftop solar panels, and electric and fuel cell vehicles would go to individuals who would have made the investment anyway. That's not much bang for the buck.

Focusing on the comparison of the CCTI with the administration's economic analysis, Congressman McIntosh challenged several agencies to defend the proposed funding for the CCTI. He submitted questions to the administration asking whether any element of the CCTI passed the Yellen Standard. A policy proposal would pass this "Yellen Standard," named in reference to Yellen's Kyoto Protocol testimony with the illustrative estimate of $23 per ton, if it cost less than $23 per ton. As evidenced in the EIA analysis, none of the proposed tax credits would pass the Yellen Standard. Although the administration proposed several versions of the CCTI over three budget cycles, Congress refused to enact any of the tax credits. The phoenixlike quality of bad policy ideas inside the Beltway applies in the case of some of these tax credits, as the Bush administration proposed similar ineffective and costly tax credits as a part of its 2002 climate change proposal.

PROMOTING EFFICIENT INTERNATIONAL EMISSIONS TRADING

"I can say that we will not accept such a limitation, period."
—Undersecretary of State Stuart Eizenstat, responding
to a question about a German proposal to restrict international
emissions trading at a 1998 congressional hearing[29]

The rules governing international emissions trading likely served as the most contentious climate policy issue between the United States and the European Union over the 1997–2000 period. The United States supported unfettered trading for several reasons. First, based on the U.S. experience with sulfur dioxide emissions trading, as well as other environmental trading programs, the United States believed that emissions trading could achieve the greatest environmental benefit for a given economic cost. Second, the United States advocated free trade in principle, and this translated to pressing for unrestricted trading in emissions allowances. Third, international emissions trading could provide the incentive for developing countries to take on emissions targets and enjoy economic benefits from selling emissions allowances into the international market. Fourth, economic analyses undertaken before, during, and after the Kyoto negotiations showed substantial cost savings from trading. For example, the AEA estimated that efficient trading among Annex I countries could reduce the marginal cost of abating greenhouse gas emissions in the United States by nearly 75 percent.

The EU opposed unrestricted trading for several possible reasons. First, it considered international emissions trading as a weakening of the Kyoto targets, because the EU claimed such trading allowed the so-called Russian "hot air"[30] to be used by other countries and thereby increase aggregate Annex I emissions above what they otherwise would have been without trading.[31] Second, some within the EU opposed trading on ideological grounds. Some left-leaning environmentalists believed that trading constituted an inappropriate "commoditizing" of the environment. Some who favored climate change policy as a mechanism to advance their social engineering goals believed that countries should comply with their targets exclusively through domestic abatement.[32] Third, the EU may have advocated restrictions on trading as a way to gain competitive advantage against countries with higher domestic abatement costs, such as Japan and the United States.

Although the United States supported international emissions trading, it was not clear whether this provision would be sacrificed as a part of a compromise deal in the Kyoto negotiations. I was surprised at the clear, concise, and explicit statement in opposition to restrictions on emissions trading made by Undersecretary Eizenstat at the March 4, 1998, hearing. At that time, the international consensus on the trading rules consisted of no more than three sentences in the Kyoto Protocol and a promise to elaborate on the rules at subsequent negotiations. The AEA would make acceptance of any restriction on trading costly in economic and political terms. Since the $23 per ton case reflected purchases of emissions allowances representing 75 percent of the effort necessary to comply with our target,[33] any restriction on trading that would have satisfied the Europeans would have forced more domestic abatement and placed higher costs on the United States. Any EU-supported restriction on trading would have made $23 per ton impossible with virtually every respected economic model of U.S. climate change mitigation, absent a creative compliance mechanism (e.g., a safety valve that capped the price on tradable permits) or renegotiation of targets (e.g., large carbon sinks). A ratification battle on a protocol with restrictions on trading could not depend on the AEA. Further, the AEA (and the nearly 300 pages of documents published in the record of the March 4, 1998, hearing) could be used to show that the costs of an agreement with trading restrictions would significantly exceed $23 per ton.

Fortunately, the administration maintained a strong position for unrestricted trading in spite of the EU's efforts to the contrary. Its analysis of the EU's May 1999 proposal to restrict both the buying and selling sides of the international emissions trading market indicated that the restrictions could bind so severely, especially on net-selling countries, that they could eliminate virtually all cost savings from trading for the United States. This proposal also provided us with an opportunity to highlight the absurd hypocrisy of the EU position. The administration evaluated the emissions transfers under the EU "bubble" with the EU proposal to restrict trading. The "bubble" provision under the Kyoto Protocol allows a group of countries to transfer emissions allowances once before the commitment period and assesses compliance on the basis of all participating countries' emissions (all those under the "bubble"). Although these transfers represented political trades of emissions, the EU decided to exclude this flexibility mechanism from its proposal that restricted international emissions trading, joint implementation, and the Clean Development Mechanism (CDM). Using the EU's own emissions inventories and forecasts, our assessment showed that 10 of the 15 EU countries' political trades would violate either the buying or selling restrictions in its own proposal (see CEA 2000, Box 7-7). The

EU proposal never gained the support of the rest of Annex I, and negotiators at the Bonn and Marrakech conferences in 2001 reached agreement on the rules for trading that excluded any quantitative restrictions on trading.

PROMOTING DEVELOPING-COUNTRY PARTICIPATION

> "In the developed world, only two people ride in a car, and yet you want us to give up riding on a bus."
> —reported response by a Chinese negotiator to a proposal for developing-country commitments at the 1997 Kyoto Conference[34]

Immediately after the Kyoto negotiations, it was obvious that the role of developing countries in climate change policy fell well short of "meaningful participation," not to mention the Byrd–Hagel resolution. The Kyoto negotiations also clearly revealed that the administration's position on "meaningful participation" did not enjoy support among developing countries. Although the developing countries, as a negotiating bloc, stridently opposed the New Zealand proposal for a Kyoto Mandate for developing-country commitments, the administration did assume in its economic analysis that some "key developing countries" would take on emissions targets and engage in international emissions trading. The analysis assumed that China, India, Korea, and Mexico adopted emissions targets for 2010. Excluding these countries from efficient international emissions trading would increase the marginal cost in the United States by 130 percent, from $23 to $54 per ton (Clinton Administration 1998, 52).

Developing countries opposed emissions commitments because they considered such commitments synonymous with constraints on economic growth. The form of the Kyoto commitments, with targets below emissions levels at the time of the Kyoto negotiations for a majority of Annex I countries, reinforced the developing countries' perception. Certainly, no developing country could adopt an emissions commitment below its current emissions level and expect this not to inhibit its economic development. Further, some countries also believed that because the industrialized countries consumed fossil energy as the basis for their development, this entitled developing countries to do the same. Finally, developing countries in general did not support emissions growth targets because of concerns about the difficulty of forecasting future economic and emissions growth.[35]

Many European countries also were lukewarm to the idea, and several opposed U.S. efforts to enlist developing countries. Some countries (and environmental NGOs) expressed concern that the United States sought developing-country commitments padded with "hot air," or, as the term evolved, "tropical hot air." If developing countries adopted "tropical hot air" targets, so their argument went, then the United States could purchase surplus emissions allowances from these countries, thereby undermining the environmental benefits of the Kyoto agreement. The EU did not like emissions growth targets set below forecast business-as-usual emissions for developing countries out of concern that a Russian-like fall in emissions would create tropical hot air. In diplomatic circles, rumors abounded that European countries actively dissuaded some developing countries that the United States had talked to about emissions commitments.

In light of the concerns of developing countries and the EU, the administration faced the challenge of how to promote more active participation by developing countries. With a public analysis premised on developing countries adopting emissions targets, this participation needed to take the form of binding first-period emissions commitments for at least some countries. As the administration worked on its economic analysis, Yellen challenged her staff to think about how to provide an incentive for developing countries to adopt emissions commitments.

The constraints on the developing-country commitment problem can be summarized as follows: to attract participation, commitments needed to be consistent with economic development; to attract the support of European countries, commitments needed to deliver real emissions abatement from business as usual; and to facilitate U.S. compliance with its commitment at a low cost, developing-country commitments needed to be of a form that would allow them to participate in international emissions trading.

A failure to comply with the first constraint would result in developing countries not joining the community of countries with emissions commitments. A failure to comply with the second constraint would result in European countries skeptical of "tropical hot air" not supporting developing countries attempting to adopt emissions commitments. Under the consensus rules governing the negotiations, opposition from European countries would prevent developing countries from taking on commitments. A failure to comply with the third constraint would result in U.S. resources, in terms of diplomatic capital and technical support, being expended with no economic benefit in terms of compliance costs. Further, international emissions trading appeared to be the only way to make the prospect of emissions commitments economically appealing to developing countries.[36]

The concerns of developing countries and the EU reflected the same fundamental aspect of emissions commitments: uncertainty over future economic growth and energy development makes the stringency of any commitment uncertain. Developing-country rhetoric revolved around what would happen if their economies grew faster than expected, a mirror image of the EU's concern over what would happen if the developing-country economies grew more slowly than expected. From their own perspectives, both groups viewed emissions commitments through worst-case-scenario lenses.

To address these concerns about uncertainty, CEA proposed developing-country emissions growth targets indexed to indicators of economic growth (see CEA 2000, Box 7-6; Lutter 2000). Such commitments would allow a country's emissions to grow from current levels as an explicit function of economic growth. Countries that experienced rapid development would have a larger emissions target; those with more modest development would have a smaller target. The key to a selecting such an indexed target lies in identifying a function that results in a reasonable below–business-as-usual emissions commitment. Constructing such an indexed target, however, should not create perverse incentives. For example, historic carbon dioxide emissions predict future emissions well. Creating an indexed emissions target based on emissions from a previous year (or years) would give countries an incentive to increase their emissions until the beginning of the first commitment period instead of implementing policies to abate emissions.

Based on our analyses on the prospects for this type of commitment, and the strong desire within the administration to break the deadlock with developing countries, CEA began introducing the idea to developing countries in spring 1998. Over the next two-plus years, we took a very active role in the administration's developing-country diplomacy at both political and staff levels. We had scores of discussions with non–Annex I countries about indexed targets in bilateral and multilateral settings. These included meetings with reluctant countries, such as China, Mexico, and Korea, and those considering more proactive actions, such as Argentina and Kazakhstan.

The first non–Annex I countries to come forward and propose to take on emissions commitments were Argentina and Kazakhstan at the 1998 climate change conference in Buenos Aires, Argentina. The administration worked with both countries, providing technical support for their efforts to develop emissions commitments. This included support for emissions inventories, economic and emissions forecasts, and evaluation of emissions abatement policies.

Kazakhstan expressed interest in joining Annex I (as had five other former Soviet republics) and adopting an emissions target similar to Russia's.[37] Although several developing countries initially put up diplomatic roadblocks under the FCCC to Kazakhstan's accession to Annex I, the parties to the 2001 conference in Marrakech, Morocco, agreed to Kazakhstan's request.

In contrast, Argentina wished to set an example of a third way for developing countries to contribute to the global effort to address climate change. At the 1999 climate change conference in Bonn, Germany, Argentina announced its intention to adopt a binding emissions target indexed to economic growth for the first commitment period if the parties to the FCCC would develop a mechanism for Argentina to participate in international emissions trading.[38] Unfortunately, although economics can inform the design of policy instruments to address concerns about uncertainty (indexing) and provide an economic incentive for participation (trading), the developing-country negotiating bloc, led by China and India, refused even to allow debate about a voluntary accession mechanism in the negotiations. Without a legal way to join international emissions trading, Argentina no longer has an economic incentive to implement policies to abate emissions. Moreover, the latest economic crisis in Argentina has most likely diverted attention away from virtually all noneconomic questions, although it may also be reducing greenhouse gas emissions.

Beyond Kazakhstan and Argentina, administration officials spoke with representatives of many other developing countries about emissions commitments, although probably none as frequently or at as high a level as China. Congress had focused on China out of concern that energy-intensive industries, and associated jobs, would relocate to China, as well as to other developing countries. Some also made the very legitimate point about the limited benefits of any climate agreement that excludes the world's second-largest emitter from policy action. Obviously, China headed the list of "key developing countries," so the administration expended significant efforts to illustrate the potential benefits to China of adopting an emissions commitment. On the president's 1998 trip to China, Yellen held several meetings with senior Chinese officials to discuss climate change policy, including the potential for indexed targets. Administration officials pressed the case in subsequent staff and political-level meetings with their Chinese counterparts. In 1999, CEA drafted an analysis of the potential gains from emissions trading for China and details about constructing an indexed emissions target.[39] The administration transmitted this report to the Chinese government at the highest level. Despite these efforts, China repeatedly refused to adopt emissions commitments and erected barriers where possible to slow U.S. efforts to enlist other developing countries.

The U.S. policy for "meaningful participation by key developing countries" did not result in a long list of developing countries coming forward to take on emissions commitments. Without the administration's economic analysis, the administration (and CEA) likely would have expended fewer resources to promote developing-country climate mitigation policies. Despite the lack of developing-country commitments, these efforts have provided benefits, some of which will likely be reaped in the future. First, involving the economic agencies helped frame discussions with developing countries more in terms of their development (which they care about) instead of the global environment (which we care about). Second, the efforts to promote developing-country policies increased the human capital in some developing countries. These countries can tap more sophisticated technical staff for work on climate mitigation policies. This may facilitate CDM projects (although it is difficult to be optimistic about the CDM). More important, this may help prepare these countries to adopt emissions commitments in the future. Third, the discussions with developing countries, even those solidly opposed to our position, such as China, allowed the U.S. government to explain how emissions trading works. At one time in the negotiating process, the developing-country bloc supported the EU call for restrictions on trading. The education and information exchange of our developing-country strategy helped weaken the resolve for such restrictions.

Finally, introducing indexed targets had two effects. It showed an innovative way for developing countries to adopt emissions commitments. It also introduced in concrete terms for the first time flexibility in the construction of an emissions commitment.[40] Although many economists support an emissions tax or a hybrid tax-trading system (sometimes referred to as the safety valve) to provide flexibility,[41] this idea faced a variety of political obstacles and gained little traction over the 1997–2000 period. Interestingly, the idea of an emissions commitment indexed to economic growth gained acceptance by the Bush administration when it proposed its alternative to the Kyoto Protocol for the United States.

THE BUSH ADMINISTRATION CLIMATE CHANGE POLICY

"When a person is accustomed to one hundred and thirty-eight in the shade, his ideas about cold weather are not valuable."
— Mark Twain[42]

In a March 13, 2001, letter to several members of the U.S. Senate, the Bush administration announced that it opposed the Kyoto Protocol.[43] The letter cited concerns about developing-country participation and the costs of the Kyoto agreement as the basis for its decision. Upon rejecting the Kyoto framework, the Bush administration convened a cabinet-level policy process to develop a new climate change policy. As the Bush administration deliberated over its policy options, the rest of the world proceeded with the international negotiations to finalize Kyoto's implementation rules. At meetings in Bonn, Germany, and Marrakech in summer and fall 2001, the rest of the world resolved its disagreements about the implementation of Kyoto. As of November 2003, 32 Annex I and 87 non–Annex I countries had ratified the Kyoto Protocol, and the agreement will enter into force if Russia (or the United States) ratifies the protocol.

Nearly one year after rejecting the Kyoto Protocol, the Bush administration announced its climate change policy. In contrast to the binding emissions quota approach embodied in the Kyoto agreement, the Bush administration proposed a greenhouse gas intensity goal. Noting that the U.S. economy currently emits about 183 metric tons of greenhouse gases (in carbon equivalent) per million dollars of output, the administration proposed a goal of reducing this emissions intensity to 151 metric tons per million dollars of output by 2012. This goal represents an 18 percent improvement in emissions intensity over the next decade. Absent any new climate or energy policies (under a business-as-usual forecast), however, the economy's emissions intensity would improve 14 percent over the next 10 years. The Bush administration's proposal represents an improvement of 4 percent in emissions intensity, or a reduction of about 100 million metric tons from what the output would be otherwise, given current economic projections.[44]

Several aspects of this policy proposal deserve mention. First, as an interesting historical comparison, the current Bush administration proposal appears to be less ambitious than the nonbinding goal accepted by President George H. W. Bush in the Framework Convention on Climate Change. The FCCC established a nonbinding goal for industrialized countries, including the United States, of stabilizing greenhouse gas emissions at their 1990 level starting in 2000. Based on the 1993 Climate Change Action Plan, attaining such a goal in the United States would have required an emissions abatement of 113 million metric tons from the 2000 business-as-usual forecast (U.S. DOS 1997, Table 4-5).[45] If the U.S. economy grew faster over the subsequent decade than projected in the early 1990s, as it did, then the country would need to undertake even more emissions abatement to comply with that goal. In contrast, the current Bush administration established a unilateral nonbinding emissions commitment that would require 106 million metric tons of abatement (based on White House–released data), and even less if the economy grows faster than expected. Promoting less stringent goals would make sense if we had learned over the past 10 years that abatement costs were exceeding previous forecasts. The very limited experience with voluntary programs, however, does not provide a basis for that claim.[46]

Second, the Bush administration has not published an analysis that illustrates how its suite of climate change policies will result in attaining the proposed intensity goal. One would prefer at least an assessment of the costs and emissions reductions from the various policies to allow for an evaluation of cost-effectiveness, if not a whole benefit–cost analysis. This could illustrate a consideration of alternatives (something congressional Republicans requested of the Clinton administration) and explain the choice of an 18 percent improvement in emissions intensity. Such an analysis could also address some concerns the public may have about the intensity goal. For example, the annual rate of improvement in carbon dioxide emissions intensity over the 1929–1998 period averaged 1.9 percent.[47] The Bush administration proposal implies a rate of improvement in greenhouse gas emissions intensity over the next decade of 1.9 percent per year. Obviously, forecasting involves more than a mere extrapolation of the past. Nevertheless, an assessment of why the next 10 years, forecast under business as usual to improve by less than 1.4 percent per year, would differ from the average of the past 70 years may better inform the public about the Bush administration's proposal and how it would require real abatement effort.

Third, complying with the greenhouse gas intensity goal could impose greater emissions reductions and higher costs in recession, and fewer emissions reductions and lower costs in an economic boom. One might design a climate mitigation policy such that if the country became wealthier, it would undertake more emissions abatement. The opposite holds for the Bush administration proposal. If the U.S. economy grew faster than expected, then the country would need to undertake less emissions abatement to attain the goal. The Bush administration used EIA's 2001 emissions and economic forecasts, noting that the EIA economic forecast matched the most recent budget's economic forecast. In addition to its reference case with 3 percent economic growth rate through 2020, EIA also conducted a forecast with a higher annual rate of 3.4 percent. I used the economic and emissions forecast from this higher economic growth case and found that faster growth *reduced* the necessary emissions abatement by 40 to 50 percent.[48] If the economy were to grow more slowly than expected, say at a rate of 2.4 percent per year over the next two decades (EIA's low economic growth case), then attaining the emissions intensity target would require about 25 to 33 percent *more* emissions abatement.[49] It does not appear to make much sense to develop a policy that would require more abatement expenditures when the country has less income than it would if the country had more income than expected. This explains in part why the Argentine government proposed an emissions target indexed to the square root of economic output. With a square root function instead of a simple linear function (the Bush administration's ratio), the emissions abatement required to comply with the target increases with economic growth. A Bush administration analysis could also explain the choice of the form of indexing and characterize potential alternatives.

Fourth, the Bush administration proposal appears to illustrate strong bipartisan support for relatively painless and ineffective policies. The proposed policies in the February 2002 announcement reflect in large part repackaged Clinton administration policies. The policies the Bush White House highlights in its policy book—tax credits, R&D funding, voluntary industry consultations and agreements, promising "credit" to firms that undertake early emissions abatement, and so on—served as the core components of the first stage of the Clinton administration climate change policy.[50] The significant difference, at least in principle, reflects the Clinton administration policy for a domestic tradable permit system that would ensure compliance with the Kyoto target. More important, this likely would help drive some of the technology development and diffusion and complement some of the voluntary actions. The Clinton administration did not develop a tradable permit system, but this reflected as much the successful efforts of congressional Republicans to forbid any work on implementing Kyoto in the executive branch as it did the lack of appetite within the administration for a potentially large policy debate.[51]

The Bush administration had the opportunity to advocate reasonable policies to promote real emissions reductions. For example, it could have supported a policy similar to the Resources for the Future economywide trading with safety valve program (Kopp et al. 1999). Some critics would have liked the administration's support for a four-pollutant utility cap-and-trade program instead of the three-pollutant program in its Clear Skies

Initiative. Instead, the climate change proposal reflects the easy short-term solution of voluntary measures. Unfortunately, the Bush administration has not provided a public analysis to inform the debate over its proposal. Although the current White House may not want to be constrained by such an analysis, good policymaking should inform the public about the economic implications of the policy in question.

CONCLUSIONS

Although some observers criticized the AEA for its optimistic assumptions about international emissions trading and developing-country participation, the focus on results premised on these assumptions yielded two benefits. First, the AEA illustrated the substantial cost savings from both international emissions trading and expanding the international tradable permit market to include developing countries. Informing the public, Congress, and foreign policymakers about the gains from emissions trading helped mobilize support for cost-effective policy implementation. Second, highlighting that "modest costs" depend on efficient international emissions trading and developing-country participation raised the political cost of failing to attain the diplomatic objectives necessary to deliver this outcome. This motivated efforts to secure good trading rules and developing-country participation.[52] Had the administration submitted the protocol to the Senate, the analysis also would have illustrated the costs of the ratification package, even if these differed from the commonly referenced $23 per ton case.

The debate over a public policy can always benefit from public economic analysis. One can always find flaws with an economic analysis: Was it a Democratic or a Republican analysis? Was it funded by the oil industry or by environmentalists? Nevertheless, it can inform the public and spur a discussion of the economic implications of the policy. Understanding the critical elements of any policy can positively influence international negotiations as well as domestic policy design. The alternative may not improve the policy debate. An imperfect analysis may be like walking in a cave with a candle: one may prefer a flashlight to a candle, but that candle sure beats walking around in the dark.

NOTES

The author thanks Randy Lutter, Ron Minsk, Jay Shogren, and Bob Cumby for productive discussions and helpful comments. All website citations below were current as of June 3, 2004.

1. Mike Toman's chapter in this volume explores this issue in greater detail.

2. Attributed to Mark Twain in the 1921 Ford Calendar (see www.twain quotes.com/Weather.html).

3. "Industrialized countries" refers to members of Annex I to the FCCC and Annex B to the Kyoto Protocol. With only a few minor exceptions, the memberships of both annexes are identical (Annex B includes 38 countries). The group of industrialized countries includes almost all Organization for Economic Cooperation and Development (OECD) countries and most economies in transition from the former Soviet bloc. All countries excluded from Annex I and Annex B are referred to as developing countries in the climate change policy context.

4. See decision 1/CP.1 in the report on actions taken at the 1995 Berlin conference at unfccc.int/resource/docs/cop1/07a01.pdf.

5. The models used in the IAT include the second generation model (SGM), the DRI macroeconometric model, Markal-Macro, and the Energy Information Administration's NEMS model. The draft report presented results primarily from the first three models. The SGM is the only global model in this set and was used to evaluate international emissions trading. The DRI model is the only one of the first three that could provide estimates of changes in unemployment.

6. The IAT evaluated three emissions targets: 1990 emissions by 2010; 1990 – 10% emissions by 2010; and 1990 + 10% emissions by 2010.

7. Two other contributors to this volume had critiqued these technology assumptions as comparable to the Great Leap Forward in an interagency memorandum.

8. See Subcommittee on Energy and Power (1998a, 75–134) for the reviewers' comments.

9. Note, however, that some economists still expressed concern about a "fast train to the wrong station" (Wiener 1999b) and preferred decision makers to pursue emissions taxes (Cooper 1998; Nordhaus 2002). Since the "t" word has been considered one of the most obscene utterances one can make inside the Beltway, policymakers outside of the economic agencies never seriously considered carbon taxes.

10. This report employed what is referred to in the literature as a "bottom-up" approach.

11. See Jacoby 1999 for a critique of this study.

12. See www.state.gov/www/global/oes/971210_clinton_climate.html for the full text of President Clinton's December 10, 1997, press statement on the Kyoto Protocol.

13. Often referred to as the "CEA analysis" because Yellen testified on behalf of the administration on the economics of the Kyoto Protocol, the AEA reflected the

contributions of multiple agencies and the administration's view of the economics of Kyoto.

14. Some in the environmental community were also concerned that the low costs implied that the United States would attempt to buy its way out of its commitment by purchasing emissions allowances from Russia and other countries in lieu of domestic abatement.

15. The remainder of this chapter will focus on the $23 per ton scenario. The $14 per ton scenario is identical except that it assumes that the European Union (EU) would not participate in international emissions trading with non-EU Annex I countries. While this possibility was under discussion during and immediately following the Kyoto negotiations, subsequent talks revealed that at least some EU countries would participate in some form of international emissions trading under the Kyoto agreement. This scenario became less relevant than the $23 per ton scenario over time.

16. See Clinton Administration 1998, Edmonds et al. 1992, and MacCracken et al. 1999 for a discussion of the SGM.

17. Although some modelers evaluated potential restrictions on trading based on the EU position on supplementarity, the modeling scenarios still assumed least-cost abatement subject to the supplementarity constraint.

18. The gains from trading presented in the AEA also reflected the inclusion of all six types of greenhouse gases. The administration evaluated the Kyoto commitments in terms of all greenhouse gases, not just carbon dioxide as had been the practice in the modeling community until then. Implicit in our modeling results is efficient intergas trading. Allowing for intergas trading reduced the price of carbon equivalent in this analysis, because many countries had much slower growth rates forecast for non–carbon dioxide greenhouse gases than for carbon dioxide. Subsequent research by the academic community substantiated this finding that accounting for all greenhouse gases can result in lower costs than a carbon dioxide–only analysis (Reilly et al. 1999; Hayhoe et al. 1999; Manne and Richels 2000).

19. The decision to maintain the EIA benchmark technology assumption confronted significant pressure while we drafted the AEA report. As the interagency discussions on the report were nearing a conclusion, the head of an environmental nongovernmental organization (NGO) wrote to Yellen, as well as other senior members of the administration, questioning the EIA benchmark technological assumption and requesting that the administration consider a more ambitious technological premise. Although such an assumption would serve some environmentalists' desires for substantial domestic emissions abatement, the more ambitious technological assumption lacked any economic justification and was rejected as the administration concluded the drafting of the report. The intervention by this environmental group also reflected a problem within the administration concerning the leaking of documents and information on internal deliberative discussions.

20. Some critics requested that the Clinton administration submit the Kyoto Protocol to the Senate in early 1998 so that it could quickly reject the agreement (see Subcommittee on Energy and Power 1998b). Although the Clinton administration did not submit the protocol to the Senate for its advice and consent to ratification, it is important to recognize that the United States was not alone. Because the negotiations about the implementation of the agreement lasted through the 2001

Bonn climate change conference, most Annex I countries had decided to wait until the details had been finalized before considering ratification. Romania was the only Annex I country to ratify the Kyoto Protocol prior to the 2001 Bonn talks. (For a list of countries that have signed or ratified the Kyoto Protocol, refer to unfccc.int/resource/kpstats.pdf.)

21. See the May 19, 2000, Joint Statement on Cooperation on Environment and Development between the United States and China, www.usembassy-china.org.cn/press/release/2000/gore519.html.

22. See, e.g., OECD 1995.

23. Quoted in Subcommittee on Energy and Power (1998a, 13).

24. The three-stage plan established the president's Climate Change Initiative over the 1997–2012 period. Stage one focused on voluntary efforts, including the CCTI, industry consultations, and credit for early action. Stage two involved a scientific and economic review and expansion of successful first-stage programs. Stage three (2008–2012) would implement the Kyoto Protocol through a domestic tradable permit system for greenhouse gases coupled with international emissions trading. See www.state.gov/www/global/global_issues/climate/background.html for more details on the Clinton plan.

25. These questions were not new; staff from economic agencies questioned the cost-effectiveness of some of these proposals during the development process within the administration. Because the administration had already decided on the size of the CCTI package, however, the goal of this policy process was to find ways to spend $5 billion or so over the next five years. This caused some to search for ways to spend the money, even if some proposals were not very effective ways to do so.

26. Congressman Sensenbrenner chaired the House Science Committee, and Congressman McIntosh chaired a subcommittee of the House Government Reform Committee. These two committees convened many hearings on climate change policy over the 1997–2000 period.

27. EIA focused primarily on the tax credit elements of the CCTI, given the difficulty in estimating the payoffs from increased R&D funding.

28. Although EIA provides a variety of measures of projected tax revenue reduction per ton of carbon abated, my discussion focuses on the scenario with forgone revenues and emissions abatement discounted at a 7 percent rate (see Table ES5 in EIA 2000a).

29. Quoted in Subcommittee on Energy and Power (1998b, 44).

30. Russian "hot air" refers to the difference between Russia's Annex B emissions commitment (1990 level) and its forecast business-as-usual emissions for the first commitment period. As a result of transitioning from central planning to a market economy, Russia's greenhouse gas emissions are expected to be below 1990 levels during the commitment period even without any new emissions abatement policies.

31. This argument suffers from two flaws. First, the opportunity to bank emissions allowances for use in subsequent commitment periods indicates that the "hot air" tons would be emitted either by a buying Annex I country in the first commitment period or by Russia in the second or a later period. Trading, even of hot air, is

then climate neutral, given that changes in the climate are driven by the stock, not the flow, of emissions. Second, the negotiations on emissions commitments were conditional in part on the negotiations on the Kyoto flexibility mechanisms (international trading, joint implementation, CDM). Some Annex I countries accepted more stringent targets with the assumption that unrestricted trading could reduce their costs more than they would have without the availability of trading. Trying to restrict trading then constituted a backdoor attempt to renegotiate the Kyoto targets.

32. Wiener (1999b, 777) notes that some referred to "these pseudo-environmentalist social engineering lobbyists" as "watermelons—green on the outside, red on the inside."

33. The 75 percent figure is from the administration tables published in the record of the March 4, 1998, hearing (Subcommittee on Energy and Power 1998b, 323). I never understood why opponents of the Kyoto agreement who tried to embarrass Yellen in hearings by trying to get her to state this figure did not just read her agency's documents. These documents were requested on a bipartisan basis, so presumably any member of Congress could have asked a colleague for them. Moreover, copies of these documents, as a part of the hearing record, were available in every government document depository in the nation by the end of 1998.

34. Reported in Climate Action Network (1997).

35. For example, in June 1998, as a member of a small delegation meeting with the government of Korea in Seoul, we raised the issue of an emissions growth target with the Korean government. Most of our meetings occurred in hot and stuffy conference rooms, which signaled the state of the Korean economy (the government shut down air-conditioning systems in its office buildings in response to the Asian financial crisis) and related challenges in forecasting economic activity. We were told that the government's forecast of economic activity over the next several quarters was highly uncertain. Given the uncertainty over forecasting the next several quarters, the Korean officials found it implausible to consider basing policy on forecasts over the next decade or two.

36. Although some of the literature on international environmental treaties considers side payments as one mechanism to promote developing-country participation, given political and budgetary constraints, such proposals were not on the table.

37. In a February 1999 meeting with the ministers for the environment and the budget of Kazakhstan, the environment minister offered this deal to the U.S. delegation: "If the U.S. gives Kazakhstan $500 million by April [1999], then Kazakhstan would give the U.S. as many emissions permits as it wants." I responded, "Unfortunately, I do not have my checkbook with me." After a stunned stare from my colleague from the State Department and a pause for the translation, fortunately followed by laughter, the environment minister repeated the deal, and I had to provide a less creative response. I am still waiting for the State Department to invite me to join the diplomatic corps.

38. The proposed Argentine target was based on this function: greenhouse gas emissions = $151.5*(GDP)1/2$.

39. Virtually every global energy-economic model shows that China would have lower marginal costs to abate a given percentage of its emissions than all other major countries in the world.

40. For further discussion of various types of developing-country commitments, see Philibert and Pershing (2001), Aldy et al. (2003a), and Bodansky (2003).

41. Proponents of such a hybrid approach include several economists formerly employed in the Clinton administration. See, e.g., Kopp et al. (1999) and Aldy et al. (2001).

42. From "Following the Equator" (see twainquotes.com/Environment.html).

43. See www.whitehouse.gov/news/releases/2001/03/20010314.html.

44. All figures in this paragraph are from the Bush administration policy briefing on the White House website (see www.whitehouse.gov/news/releases/2002/02/climatechange.html).

45. This reflects emissions of carbon dioxide, methane, nitrous oxide, and synthetic compounds. It does not include estimates of carbon sequestration to facilitate the policy comparison. Note that the 2002 climate change policy reflects gross emissions and does not account for carbon sequestration (see www.whitehouse.gov/news/releases/2002/02/addendum.pdf).

46. Proponents of these voluntary programs would claim to the contrary that emissions abatement costs less than some may have previously predicted. I think these programs do little more than subsidize corporate public relations departments and illustrate the phenomenon of adverse selection to economics students.

47. This average reflects U.S. economic output data from the Bureau of Economic Analysis and carbon dioxide emissions from the Carbon Dioxide Information Analysis Center. EIA carbon dioxide emissions data date back to 1949; emissions data for other greenhouse gases date back only to 1980.

48. EIA publishes forecasts only for energy-related carbon dioxide emissions and GDP. My range reflects two assumptions. The low end assumes that the increase in non–carbon dioxide emissions is proportional to the increase in carbon dioxide emissions in the higher GDP growth case. Since non–carbon dioxide emissions tend to track economic output less closely than carbon dioxide, the high end of the range assumes that forecast non–carbon dioxide emissions are identical under the reference and high economic growth cases. See Appendix B of EIA 2000a for the high and low economic growth cases.

49. This calculation is identical to the calculation describe in the preceding footnote. The high end of the range assumes that non–carbon dioxide emissions are identical under the reference and low economic growth cases, and the low end assumes that non–carbon dioxide emissions grow more slowly in the low economic growth case than in the reference case.

50. Interested readers may want to compare the informational materials released by the Clinton administration in 1997 on its climate change policy with the Bush administration informational materials. Clinton administration: www.state.gov/www/global/global_issues/climate/background.html. Bush administration: www.whitehouse.gov/news/releases/2002/02/climatechange.html.

51. Congress attached to several years' appropriations bills the Knowlenberg Amendment, which strictly forbade the staff at many of the relevant agencies to work on any policy related to implementing the Kyoto Protocol. Because the Senate requested implementation legislation to accompany the Kyoto Protocol when the administration decided to submit the agreement to the Senate for its advice and

consent (under the Byrd–Hagel resolution), one may wonder how the administration could have satisfied these conflicting preferences.

52. Two years after we conducted the AEA, it had become evident that China, India, Korea, and Mexico were not going to adopt emissions targets for the first commitment period. Although the climate change issue had a lower Beltway profile at this time, the concern about costs and "insufficient" domestic emissions abatement (environmentalists do not give up easily) prompted some to suggest that we redo the AEA in 2000. In this case, the proponent of this idea wanted the new analysis to assume 100 million metric tons of emissions reductions for free as a result of the president's announced goal to triple U.S. biomass energy by 2010. The "policy" supporting this goal was an executive order convening some new government committees.

CHAPTER FIVE

MAKING MARKETS FOR GLOBAL FOREST CONSERVATION

Jonathan B. Wiener

T his is a story about forests, in the past and in the future. Many of our stories about forests, at least in Western civilization, depict the forest as dark and evil. In the tale of Little Red Riding Hood, the forest and its wolf represent the Bad; the girl and her grandma the Good; and the day is saved by the woodcutter, whose job is to cut down the forest.

My story views the forest more hopefully. The world's forests are a global public good. They harbor biological diversity that supports ecosystems and promises new biotechnologies, and they sequester carbon that would otherwise heat the atmosphere. Conserving forests yields environmental and economic benefits.

This is also a story about the Council of Economic Advisers (CEA), in its own way a small stand of trees amidst the vast forest of American government. It is a story of my year and a half at CEA, from mid-1992 to late 1993, in both the first Bush and the Clinton administrations, and my efforts to launch the Forests for the Future Initiative.

Many stories about CEA, and about economists generally, depict them as dark and evil enemies of environmental protection. This is a misunderstanding. It is true that CEA's role is often to clarify the costs of environmental regulations and the flaws in policy designs. Indeed, the title of this book is a bit ironic given that the environmental economists at CEA are sometimes obliged to *prevent* the White House from being painted too green, at least in the sense of environmentalism. More generally, a major function of CEA—which has no specific statutory responsibilities other than to publish an annual report and to give advice—has been to stop bad ideas.[1] That is usually because CEA has no special constituency; it is a bulwark of sober analysis against the distortionary pressures of organized special interest groups.[2] Hence many interest groups, including but hardly limited to environmentalists, may see CEA as an obstacle, at least to parochial and poorly conceived policies. Rightly so. CEA in this role is not the enemy of environmental protection; it is the enemy of narrow and counterproductive policies. It is quite favorable to proposed environmental regulations that harness incentives to achieve net benefits; prime examples include the phaseout of lead in gasoline and the phaseout of CFCs.

My story sees CEA in its more "positive, creative role."[3] Because CEA sees issues broadly, it can innovate policy solutions that yield collective benefits—policies that individual interest groups would not have had the incentive to develop. CEA can be a public goods entrepreneur. It can practice Schumpeterian innovation, but in the public sector rather than in the private sector setting that Schumpeter described: it can invent new institutions to produce the shared public goods that conventional competition among special interest–oriented politicians and government agencies would not. In these cases, CEA can promote effective and efficient environmental protection. My story is one example. It is the story of CEA trying to create a new institution to conserve the world's forests.

COSTS AND BENEFITS OF FOREST CONSERVATION

In 1992, when I came to the White House,[4] the Rio Earth Summit was impending. Climate and biodiversity were on the agenda, but forests were not getting much attention. Forest loss was a serious and immediate problem: from 1980 to 1990, the forests in poorer countries were being cleared at about 13.5 million hectares (0.7 percent) per year, while the forests in wealthier countries were expanding at about 1.1 million hectares (0.1 per-

Table 5-1. Rate of Change of Forest Cover

Country group	Forest cover (mha)			Average annual change (%)	
	1980	1990	1995	1980–1990	1990–1995
"Developing"	1,930	1,796	1,733	-0.7	-0.7
"Developed"	1,704	1,713	1,720	0.1	0.1
Total	3,634	3,511	3,454	-0.35	-0.33

Sources: FAO (1999, 1-135, Annex 3, Table 3); WRI (1998, 292-93, Table 11.1).

cent) per year, for a net loss of about 12.4 million hectares (0.35 percent) per year.[5] (See Table 5-1.) These rates alone do not reveal whether they are acceptable or excessive. And forest loss is not just a recent problem; humans have been clearing forests since long before the industrial revolution.[6] But there are good reasons to think that current forest losses are inefficiently high.

First, a systematic mismatch of costs and benefits leads to overclearing (underconservation). Stated as a problem of externality, the full social and environmental costs of forest losses are not borne by those who convert forest lands to nonforest uses. The consequence is excessive forest loss and degradation—excessive in the sense that if they faced the full costs of their forest conversion, forest users would find it in their own interest to conserve more standing forest and convert less forest to other land uses, and the world would be better off as a result. Stated as a problem of public goods, the shared benefits of forest conservation are not reaped by those who could conserve them on the ground. That is because the global benefits of forest conservation accrue to people far removed from the locations at which decisions to conserve or convert forests are made, and accrue outside market transactions. These global benefits include the value of biodiversity to ecosystem vitality, the value of carbon sequestration to preventing global warming, and the value of both of these to future as well as present generations. No individual has an adequate incentive to invest in providing this collective good, because each knows that he or she will benefit as long as others do the job; each can take a "free ride" on conservation efforts by others. In short, individual nations and private actors would bear costs to keep forests standing but would reap only a fraction of the global benefits. Hence forests continue to be cleared, and biodiversity and carbon storage continue to be squandered.[7]

To be sure, conserving forests yields some local benefits that are (or can be) internalized in local decisions. It is more accurate to speak of at least three concentric circles of forest conservation effects, three ripples in the pond of incentives: local, national, and global.

The first circle represents the benefits and costs borne by the local land user. These include the value of the timber and other products harvested for consumptive use and the local effects of clearing, such as reduced shade and accelerated soil erosion. Most land users will internalize and optimize these effects.[8] Even where local externalities do occur, transaction costs may be low enough to permit beneficiaries to pay holders to internalize and conserve,[9] such as via ecotourism and contracts to screen genetic resources for pharmaceutical research.

A second circle of effects is national: the benefits and costs that go beyond the local land user but are contained within the nation-state. These might include the costs of landslides on steep slopes, regional ecosystem disruptions, and regional changes in albedo and transpiration that disrupt weather and precipitation. They would include the collective benefits to the society of conserving forests for future generations. These costs and benefits would be treated as externalities by the local land users, who would free ride on others' efforts to conserve forests for such reasons. In this case, the transaction costs of deals between forest owners and their numerous beneficiaries might be frustratingly high. One might expect national governments to adopt corrective policies to force land users to internalize such costs and benefits in the interest of national efficiency, but national governments tend to promote excessive forest clearing rather than prevent it.[10] The causes of such mismanagement are complex but essentially involve the parochial motives of subparts of the bureaucracy to expand their own budgets or domains at collective expense. Nonetheless, there is at least a thin hope that over time, national political systems will reform to rectify these situations as they become too costly to the society at large.[11]

The third and largest circle is global: neither local land users nor individual national governments can be expected to internalize these costs and benefits. These include overall biodiversity conservation, especially as an insurance strategy for life on earth as a whole, and carbon sequestration to forestall global warming. These costs and benefits will be treated as external to the nation, and nations will free ride—consume their own forests— rather than try to produce global public goods that benefit others at significant cost (in forgone uses of forest land) to themselves.[12] The transaction costs of connecting global beneficiaries to forest holders, and of binding free riders, are probably prohibitive. Hence the basic problem is one of underprovision of global public goods.

A second general reason makes the problem even more dire: the costs and benefits of the global public goods are not uniformly distributed across countries. The costs of forgoing forest clearing would fall largely on the countries experiencing rapid forest clearing. But the countries in which forest loss is rapid tend to be poorer, and therefore even less interested in conserving forests than their counterparts in wealthier countries. This is in part because the taste for environmental quality rises with wealth. Poorer countries typically perceive more urgent priorities than conserving forests, such as feeding, housing, and educating their needy populations. Thus, simply asking poorer countries to stop clearing their forests is ineffective, because those countries are unlikely to agree to bear the costs of forgoing forest conversion when the global benefits of forest conservation go largely or completely to others and to the future.[13]

Meanwhile, even though the global benefits of forest conservation are nonexcludable public goods, such as preventing global warming, they are disproportionately reaped by wealthier countries that place a high value on such goods. This is true even if poorer countries may be more physically vulnerable to such risks as global warming than are wealthier countries, because the difference in valuation as a function of wealth can dominate the difference in physical losses. For example, even if Bangladesh were physically more vulnerable to sea level rise than the Netherlands over the next century, relatively wealthy Netherlands might care much more about forestalling global warming than does Bangladesh because poorer Bangladesh is much more focused on saving itself from abject poverty in the next 10 years. This is, of course, an empirical issue, but to the extent that the global benefits of forest conservation are valued more with increasing wealth, and the costs of conservation are borne in one place while the benefits are borne in another, the prospects for internalization are even more remote than in a standard one-society problem of public goods and collective action.

A third reason for the current predicament is a fundamental feature of international law. Even if forest conservation is an important public good, and even if its beneficiaries diverge from its cost bearers, one might imagine that the relevant community could still accomplish successful collective action to yield overall social net benefits. Some benevolent despot or majority coalition of interest groups could impose the policies necessary to protect forests at the efficient level. Efficiency in this sense would be Kaldor–Hicks or Potential Pareto efficiency, in which real costs may be imposed on dissenters as long as the aggregate net benefits are positive. If such policies did impose costs on the poor to benefit the wealthy, that blow could be softened by redistributive fiscal policies (but such redistribution would not be necessary for the imposition of the efficient regulation in the first place). This is indeed what could happen in a well-functioning national democracy or a benign dictatorship. But it is unrealistic at the international level. International treaty law operates on the basis of consent, not majority rule. National legislation is binding on dissenters (apart from civil rights exceptions), but treaties cannot bind any nation unless that nation assents to the treaty. Any country that feels itself disadvantaged by joining a treaty simply need not join. In consequence, international treaties cannot impose Kaldor–Hicks efficiency (aggregate net benefits, with some parties bearing real costs). International treaties must be actually Pareto-improving for all participants: every nation must perceive net gains from joining the treaty. Those countries that would be burdened by a requirement to conserve a global public good will not agree to do so unless it is made worth their while—that is, unless they are paid by the beneficiaries.

Notice how different this is from the typical conception at the national level. At the national level, we think of forcing (coercing) sources of externalities to internalize the social costs of their actions. The catchphrase is "polluters pay." But at the international level, there is no supervening state authority to undertake such coercion. Sources of externalities will not agree to internalize their social costs unless they come out better off as a result. Hence, at the international level, because of the voluntary assent voting rule, the solution to global externalities must lie in an arrangement in which the beneficiaries of controlling the global externality (that is, of producing the global public good) finance such efforts. Thus, international environmental law must operate as "beneficiaries pay."[14] Because the beneficiaries of global environmental public goods tend to be the wealthier countries, this has comfortable implications for distributional equity. But it also means that international forest conservation efforts cannot succeed by demanding that poor forest-holding countries snap to it and stop deforesting or act sustainably. Instead, the wealthier (beneficiary) countries must find a way to finance forest conservation in the poorer (nonbeneficiary) countries.

The solution to this problem of underprovided global public goods is to create an institutional arrangement for global contributions to the financing of local forest conservation, reducing the transaction costs that currently obstruct deals between forest beneficiaries and forest holders. The key question is how to establish and structure such a system so that it furnishes efficient incentives to forest holders to provide globally shared forest benefits.

THE INADEQUACY OF CURRENT EFFORTS

Present international legal regimes do not fit the bill. The second clause of Principle 21 of the Stockholm Declaration, for example, prohibits states from taking actions that would damage the environment outside their borders.[15] That edict may cover forest uses that cause direct external damages, such as erosion and flooding in neighboring states, but it does not address a state's failure to conserve the environment within the state's borders for the benefit of the broader global community. Indeed, countries reluctant to agree to conserve their forests will likely cite Stockholm 21's first clause, which guarantees nations the sovereign right to exploit their own natural resources.

The Convention on International Trade in Endangered Species (CITES)[16] attempts to block trade in rare species, and hence to discourage hunting of such species. But CITES may provide some perverse disincentives to conservation of forests and wildlife. Its main instrument is a prohibition on trade in endangered species for commercial purposes. This may reduce overhunting of the species, but it may exacerbate habitat loss. Local holders of habitat for endangered species may find that habitat to be of less value than if they could harvest some members of the species for economic use. Thus, they may be more likely to convert the land to uses other than wildlife habitat, such as agriculture.[17]

The 1992 conventions on climate change and biological diversity may help a bit. The Framework Convention on Climate Change (FCCC)[18] obliges industrialized parties to take actions with the aim of reducing the future growth of net greenhouse gas emissions, including both reducing emissions sources and protecting and enhancing sinks (such as forests, which store carbon) (Article 4.2). Furthermore, the FCCC authorizes parties to implement that commitment jointly with other parties (Article 4.2(b)), giving parties seeking to protect forest sinks an incentive to support forest conservation efforts worldwide as well as domestically. Several pilot bilateral efforts to conserve forests for carbon storage have been undertaken under the rubric of such joint implementation. For example, AES, a U.S. independent electricity producer, sponsored a project in Guatemala to plant degraded land with new trees; NEES, a U.S. electric utility, supported a project in Malaysia to reduce the collateral damage done by selective logging to surrounding trees and thereby conserve biomass and carbon in the forest; and American Electric Power Company, Pacific Corp, and British Petroleum, together with Fundacion Amigos de la Naturaleza, the Nature Conservancy, and the government of Bolivia, launched the Noel Kempff Mercado Climate Action Plan, a carbon sequestration project that received the 2003 Roy Family Award from Harvard. Such forest conservation projects are only partially endorsed by the Kyoto Protocol to the FCCC,[19] which limits total sink enhancement by industrialized countries, fails to provide credit for conservation of existing forests, and limits sale of credits for sink enhancement by developing countries under the Clean Development Mechanism (CDM).[20]

Meanwhile, the Convention on Biological Diversity[21] may help clarify the property rights to genetic resources in the forest. Holders of biodiverse areas would then have a greater economic stake in conserving their biodiversity to license the prospecting and use of such genetic resources.[22] This institutional arrangement could translate some of the currently shared global benefits of conserving forests into local gains (and foreign investment support) to keep the forests standing.

But a more direct approach to global forest conservation is still needed. The FCCC provides only an initial push toward international forest protection, and some developing countries are resisting the option of forest conservation under joint implementation or the CDM. And, apart from its rhetoric, the main practical effect of the Convention on Biological Diversity may be to confer value on the appropriable commercial uses of forest genetic resources, without generating economic incentives to conserve the ecological functions and existence values of forest biodiversity, such as habitat for noncommercial but ecologically valuable species. Moreover, parts of the Biodiversity Convention, such as Article 16, may unintentionally work to discourage international investment in genetic resource conservation by taxing such investments. The full value of the global services of standing forests will still be incompletely incorporated into the incentives offered to local holders.

Older efforts, such as the Tropical Forestry Action Plan (TFAP), have foundered. The Non-Binding Statement of Forest Principles adopted at Rio in 1992 did little or nothing. More recent efforts, such as the World Commission on Forests[23] and the Intergovernmental Panel on Forests chartered by the UN Commission on Sustainable Development (CSD),[24] have generated useful information but have so far done little to mobilize real investments in forest conservation. A subsequent initiative, the Intergovernmental Forum on Forests, established under the CSD in 1997, was aimed at supporting a new international treaty on forests but generated little momentum.[25]

Some suggest that the remedy is to require sustainable management of forests worldwide.[26] But the notion of an international treaty or informal consensus committing all countries to adopt sustainable management of forests is illusory. It amounts to asking the poorer forest-holding countries to undertake forest conservation at a cost to them to provide global public goods. This they will not do. Such a treaty would not be Pareto-improving. Under international law, they will decline to assent. To gain the assent of poorer forest-holding countries, such a treaty would have to involve some kind of mechanism for wealthier countries to finance forest conservation in the poorer countries.

One way to get around the problem of consent might be to enforce a sustainable management requirement by deploying trade barriers such as border taxes on imported timber that was not harvested in accordance with the requirement.[27] This remedy has several flaws. First, timber harvesting (logging) accounts for only a small fraction of total forest losses. Agriculture (shifting cultivation) accounted for about 45 percent of recent deforestation in the tropics, driven by poverty and government policies encouraging land clearing.[28] In Central and South America, ranching is an important second cause; in Africa, it is fuelwood demand.[29] Logging for timber is a far less important cause, perhaps accounting for less than one-fifth of 1 percent of all tropical forest loss.[30] Thus, taxing timber products in international trade may not significantly affect the major causes of forest loss.[31] This also explains why efforts such as the International Tropical Timber Agreement (ITTA) and forest products certification programs, which address trade in timber products, are not really germane to the problem of global forest conservation.

Second, it may be difficult to fashion a border tax on timber and wood products that is both effective at conserving forests and legal according to the General Agreement on Tariffs and Trade (GATT) and the World Trade Organization (WTO). To make a border tax a GATT-legal restriction on the product and not a GATT-illegal restriction on the production and process methods (PPM), the taxing country would have to point to features of the wood as a product, such as the quality of fibers, that make it possible to distinguish unsustainably harvested from sustainably harvested trees. Although these features may be easily discernible for raw timber, the monitoring costs might be prohibitive if the fiber test had to be applied to finished wood products. Moreover, taxing imports of old-growth wood more than second-growth plantation wood does not directly induce the conservation of forests for biodiversity, carbon storage, and indigenous cultures, because those factors inescapably depend more on PPM issues: how the trees were removed (including how much damage was done to surrounding trees) and what was done with the land afterward. Making the tax GATT-legal may distort its application so much that it no longer serves its purpose. Indeed, a tax favoring plantation trees in international trade could induce forest holders to clear their old-growth forests (using those trees locally) to plant new plantations for international sales—the opposite of forest conservation for biodiversity objectives.[32] Or taxing wood products might reduce the returns to keeping land as forest compared with other land uses, and thus encourage conversion to nonforest land uses.

The basic problem with the border tax idea is that the forest services currently being underprovided, such as biodiversity and carbon storage, are not the same kind of goods as the traditional forest products, such as timber and wood pulp. Taxing the latter may not yield the former. The problem is conserving forests as a land use, not improving forestry per se. Moreover, using policy tools such as border taxes to coerce poorer countries into forest conservation may be self-defeating. Poorer countries would resist agreeing to such a regime and hence raise GATT–WTO challenges; such taxes might impair global trade more than they improve the global environment; to be effective, such taxes might need to be so high that they would not be credible (because they hurt the imposing country's economy too); even if effective, such taxes might be deemed unfair; and they might turn out to hinder compliance in poor countries by making them even poorer (which was, after all, a key reason for their rapid forest clearing in the first place).[33]

MODELS FOR A GLOBAL FOREST CONSERVATION AGREEMENT

A more direct approach would take the market failure by the horns: construct an international legal regime in which the beneficiaries of global forest services would contribute to the conservation of forests. This is what we tried to do while I was at CEA, the Office of Science and Technology Policy (OSTP), and the Justice Department just before I came to CEA.

The traditional model of a forest treaty has been national responsibility, calling on all countries to accept responsibility to conserve their own forests—that is, to halt domestic forest loss or follow specific forest management practices. The forest-holding countries will understandably resent and resist such a demand to restrict the use of their valuable resources for the benefit of all. This would be true even if they were not financially desperate, but the fact that many are compounds the predicament.

A second model might be called the "holders pay" model, akin to taxing negative externalities. The treaty would put the obligation on forest holders to pay a tax equal to the external costs imposed by forest loss. Thus, for each hectare of each type of forest cleared, there would be a tax rate. The tax would internalize the global cost (reduction in global benefits) occasioned by forest loss. This model also has serious drawbacks. Central determination of the taxes owed according to local forest loss rates, and collection of the taxes, would require an enormous monitoring and administrative apparatus. Collecting this revenue in one place would raise fears of abuse and incompetence in its disbursement. National collection would invite game playing within national tax and subsidy systems.[34] Most important, countries with rapid rates of forest loss would predictably refuse to consent to this approach.

A third model for a global forest conservation agreement is the "beneficiaries pay" model, seeking to get the beneficiaries of forest conservation to help invest in the provision of these benefits. The key role a forest treaty can play is thus to bring together forest beneficiaries and forest holders in a global arrangement. Beneficiaries would agree to help finance the conservation of global forest values, in return for such action by forest holders and a verification–monitoring system. Such a convention would explicitly state the obligation of beneficiaries of global benefits of forest services to contribute to the financing of such services and benefits.[35]

This obligation would need to be made operational. There are two basic ways to do so: as a price (an amount of money to be invested) or as a quantity (an amount of forests to protect). Our first effort in 1991–1992 was to design a quantity-based regime. This choice was partly based on our experience with tradable permits for pollution control and fisheries management. These tradable permit systems had achieved substantial environmental protection gains at a fraction of the cost of traditional regulatory approaches.[36] And the administration had just worked with the Environmental Defense Fund (EDF) to create the tradable permits system for SO_2 emissions adopted to control acid rain in the 1990 Clean Air Act Amendments.[37] But our choice of a quantity-based regime for the international level also derived from our suspicion that a price-based approach—aid, not trade—would mean subsidizing forest conservation with no quantitative goals, an approach that, as with other subsidies, can yield perverse results (for example, if the subsidy reduces forest clearing by each actor but increases the size of the forest-clearing sector of the economy).[38]

So we sketched out the quantity-based, trade-based approach. First, the amount of obligatory action would need to be agreed upon. This would be determined by the optimal amount of forest services to be provided, considering benefits and costs. As a proxy, it could be defined as an optimal amount of forest cover (including all types of forest: tropical, temperate, and boreal); and this quantity of forest cover should be adjusted for the quality of the different forest types (in terms of biodiversity value, carbon storage, and so on). A quality index of different forest types could be developed, similar in concept to the index of relative global warming effects devised by the Intergovernmental Panel on Climate Change (IPCC) to compare greenhouse gases.

Second, the global obligation would need to be allocated among states according to some acceptable formula.[39] If global benefits are correlated with national wealth or income, the obligation could be allocated according to each country's share of global gross domestic product (GDP). To take a simplified example of this scheme: assume that the global optimum would be to conserve today's total of quality-adjusted units of forest cover, which at present rates are being lost at roughly 1 percent per year. Then imagine a treaty that allocates such forest conservation obligations to all countries by share of global income, not by current forest holdings. Take the case of two countries: Q is relatively cash-rich but forest-poor, whereas Z is relatively cash-poor but forest-rich. If country Q has 12 percent of global GDP, it would be assigned 12 percent of the quality-adjusted forest conservation obligations. If Q has 5 percent of the quality-adjusted global forests within its borders (which it was committed to maintain), then Q would have to go looking for the remaining 7 percent to conserve elsewhere to make up its obligation. Country Z, with only 2 percent of global GDP but 8 percent of the world's quality-adjusted global forests within its borders, would have 6 percent extra forest cover to lease. In partial satisfaction of its obligation, Q could arrange to pay Z for Z's agreement to conserve forests in Z. (To make up Q's remaining 1 percent, Q could find another partner or make efforts to expand its domestic forest cover.)

Forest cover and GDP data for actual countries are shown in Table 5-2. (These numbers are unadjusted for differences in quality across forest types; adjusting for biodiversity value might raise the relative value of tropical forest areas over temperate and boreal forest areas.) The last two columns of this table illustrate the degree of extra financing that some

Table 5-2. Forest Cover and GDP, 1995

(1)	(2)	(3)	(4)	(5)	(6)	(7)
Russian Federation	763.50	22.1	1.2	42.96	0.06	-720.54
Brazil	551.14	16.0	2.5	85.35	0.15	-465.79
Canada	244.57	7.1	2.0	70.57	0.29	-174.00
USA	212.52	6.2	25.0	862.32	4.06	649.80
China	133.32	3.9	2.5	86.53	0.65	-46.79
Indonesia	109.79	3.2	0.7	24.57	0.22	-85.22
Peru	67.56	2.0	0.2	7.12	0.11	-60.44
India	65.01	1.9	1.2	42.43	0.65	-22.57
Mexico	55.39	1.6	0.9	31.01	0.56	-24.37
Colombia	52.99	1.5	0.3	9.44	0.18	-43.55
Bolivia	48.31	1.4	0.02	0.76	0.02	-47.55
Sudan	41.61	1.2	0.02	0.74	0.02	-40.87
Australia	40.91	1.2	1.3	43.26	1.06	2.35
Papua New Guinea	36.94	1.2	0.02	0.61	0.01	-36.33
Argentina	33.94	1.0	1.0	34.86	1.03	0.92
Tanzania	32.51	0.9	0.01	0.45	0.01	-32.06
Japan	25.15	0.7	18.3	633.65	25.20	608.51
Sweden	24.43	0.7	0.8	28.36	1.16	3.94
Angola	22.20	0.6	0.01	0.46	0.02	-21.74
France	15.03	0.4	5.5	190.53	12.67	175.50
Germany	10.74	0.3	8.7	299.65	27.90	288.91
Italy	6.50	0.2	3.9	134.82	20.76	128.33
UK	2.39	0.1	4.0	137.16	57.40	134.77
Israel	0.10	0.0030	0.3	11.41	111.85	11.31
Saudi Arabia	0.02	0.0006	0.5	15.57	707.67	15.54
Kuwait	0.005	0.0001	0.1	3.31	661.20	3.30
World	3,454.00	100.00	100.00	--	--	--

Source: Forest cover and GDP data in columns (2) and (3) are from WRI (1998, *236–37*, Table 6.1; *292–93*, Table 11.1).
(1) Country
(2) Domestic forest cover (mha)
(3) % Share of global forest cover = (2) / 3,454
(4) % Share of global GDP
(5) GDP-based share of global forest conservation obligation = (4) times 3,454 (mha)
(6) Ratio of GDP share to forest cover share = (4) / (3)
(7) GDP-based share of global forest conservation obligation minus domestic forest cover = (5) – (2) (mha)

countries would have to undertake to satisfy their GDP-based share of the global conservation effort. The next-to-last column shows the ratio of countries' GDP shares to their domestic forest holding shares; the last column shows the absolute number of hectares each country would have to conserve to match its GDP-based share of the global conservation effort (positive values imply an obligation to conserve that much more forest; negative values imply an excess of domestic forest over the GDP-based share of global conservation effort and hence an opportunity to sell or lease conservation credits). For example, the United States, Japan, and Germany would need to make major investments in expanding their domestic forests or, more likely, obtaining forest conservation in other countries. These last two columns also show the countries that would be large net recipients of forest conservation financing, such as Russia, Brazil, Canada, and Indonesia.

Third, the mechanism for translating these allocated obligations into cross-national investments in forest conservation would need to be established. The best option would most likely be a decentralized market-based system, imposing a requirement to meet one's allocated obligation through whatever mix of bilateral investments each country chooses. This would be similar to tradable credits for pollution reduction, but here it would involve tradable obligations for forest conservation—a method that gets financing devoted to the best, most effective projects.[40]

The funding for forest conservation would go to various projects and programs in forest-holding countries. It could go to governments or non-governmental organizations (NGOs) or a combination thereof. Each activity would be measured by its ability to generate global benefits of forest conservation and be monitored over time. Thus, under the decentralized system, each country would need to invest in conservation efforts that would achieve its obligated level of forest conservation.

Wealthier countries with obligations exceeding their domestic forest holdings would compete to find the best forest conservation ventures, and poorer countries with forests exceeding their obligations would compete to attract the best investments. Poorer countries with forests would now have a major incentive to conserve those forests; they would not be asked to conserve forests at their own expense for others' benefits. A corollary result would be a stimulus to innovation in efforts to conserve forests; entrepreneurial forest conservation would become a profit center for poor countries instead of a liability.

Would poorer countries participate? The main attraction would be the profits they could earn from selling or leasing forest conservation credits to wealthier countries at prices exceeding their costs of conservation. Some forest-holding countries might fear losing their sovereign control over land assigned to conservation financiers in wealthy countries. But this should not be a real fear; countries need not agree to such transactions unless the price is high enough to exceed their opportunity costs, and countries could lease forested lands instead of selling them outright. Still, this emphasizes that a global forest conservation treaty should address global public goods only and not become entangled with local land ownership. A related possibility is that under such a scheme, some poorer countries might hold out to try to extort high prices from the wealthy countries. If the market is large enough, competition should alleviate this problem, as holdouts would know that they could get left holding the bag when investors signed with others. Further, the option to satisfy the obligation via domestic afforestation (presumably at a fairly low quality-adjusted rating per hectare) would enable wealthy countries to expand the market, sidestepping holdouts and putting an upper bound on the cost of compliance.

Would the wealthy countries enter into this scheme? Only if they see the value of conserving the forests as exceeding the cost, both collectively and individually; otherwise they will let the forests disappear. And no one country would do this on its own, because it would then bear the costs while the benefits would be shared collectively. So it takes a collective agreement, with monitoring to prevent defections. Reducing the cost of participation, through a cost-effective system of tradable forest conservation obligations, would increase the likelihood of participation by wealthy countries. Still, some beneficiary countries that would have large outstanding obligations, such as countries with large GDPs relative to their domestic forest covers—consider the net obligations of the United States, Japan, Germany, France, Italy, and the United Kingdom, shown in the last column of Table 5-2—might refuse to sign the convention. A key to the treaty's success would be obtaining a large fraction of both beneficiaries and forest holders.[41]

A longer-term problem is that forest conservation takes time, and the well-intentioned efforts by beneficiaries might turn out not to succeed in reducing forest loss. Some international administrative apparatus would be needed to monitor results, such as in terms of quality-adjusted forest cover. This problem confronts every effort to encourage forest conservation; it is not unique to the system of tradable conservation obligations sketched here. The cost savings enabled by a trading system could be expected to reduce noncompliance and far exceed any added costs of monitoring. Moreover, a system of explicit obligations would focus greater attention on monitoring and achieving successful forest conservation.

By contrast, a "tax and pay" model would tax forest holders for damages to forest values, while requiring the beneficiaries to pay the forest holders side payments necessary to engage the holders' consent to be taxed. The problem with this approach is that to secure consent, the side payments must vary in proportion to the holders' net costs of joining the treaty, and hence in proportion to the holders' net forest conservation. This means that the side payments will undermine the incentive effect of the tax they are relieving. And because the essential feature of a tax, as contrasted with a quantity instrument, is that the tax fixes the cost but does not fix the quantity of abatement or conservation,[42] the ability of the tax to achieve global forest conservation would be seriously undermined if not offset.[43]

THE FORESTS FOR THE FUTURE INITIATIVE

While I was at the Justice Department and OSTP, just before coming to CEA, I worked with colleagues at several offices (chiefly the White House Counsel's Office and EPA) to develop these ideas. We retained (through EPA) a leading forest economist, Roger Sedjo of Resources for the Future (RFF), to develop a model that would flesh out the proposal. Sedjo developed an index of forest quality and calculated the financial flows that would occur across countries as a result of the GDP-weighted tradable forest conservation obligation scheme.

In 1992, we met opposition from two powerful camps: the Office of Management and Budget (OMB) and the State Department. OMB and State had collaborated on the U.S. position on the Non-Binding Statement on Forest Principles to be adopted in Rio, urging sustainable management by all countries. State's main objection was that rolling out a new initiative on forests would shake things up. "No new initiatives," said State. Our response was that the U.S. position in the run-up to Rio was in sad shape, at least in the public eye (opposing targets in the climate treaty and ultimately deciding not to sign the biodiversity treaty), and that a new proenvironment initiative was just what the United States needed. Ultimately, the State Department agreed to let the initiative be announced, but only on the day before the Rio conference opened.

OMB objected to a quantity-based regime precisely because it would fix the quantity obligated and leave the cost of doing so uncertain, to be set by market forces.[44] OMB did not want the U.S. budget to be on the hook for an uncertain and possibly escalating price.

The compromise was to transform the quantity-based tradable forest obligation proposal into a price-based offer of a specified investment by the United States and make an effort to get other wealthy countries to make similar pledges. Thus, on June 1, 1992, President George H.W. Bush announced the Forests for the Future Initiative (FFI), pledging to invest "$150 million . . . as the US downpayment toward a worldwide doubling of international forest assistance."[45] This new initiative was cochaired by William Reilly, administrator of U.S. EPA and the head of the U.S. delegation at Rio, and Boyden Gray, the White House counsel. Several agency officials and I acted as the staff of the initiative. The funds were to be offered through a Request for Proposals, so that instead of dictating to poorer countries how the United States thought they should manage their forests, the United States would solicit competitive bids from countries (and NGOs) reflecting the bidders' own choices of their best forest conservation projects.

We went to Rio to promote FFI and to shop it with other countries, both wealthy and poor. We found several eager country partners. The ambassador of Romania gave a spontaneous speech at Rio praising FFI. Several NGOs, notably the Wilderness Society and the World Wildlife Fund (WWF), publicly endorsed FFI.

When I returned from Rio, I began my stint at CEA, continuing to work on many of the same projects that I had begun in concert with CEA while I was at OSTP and the Justice Department, including FFI. That summer and fall, I helped run an interagency team that solicited and reviewed proposals and negotiated start-up activities with interested countries and NGOs.

In a January 1993 ceremony in the Roosevelt Room of the White House, President Bush formally launched a set of eight initial partnerships under FFI—with Belize, Brazil, Ghana, Guatemala, Indonesia, Mexico, Papua New Guinea, and Russia. Several NGOs were involved, including WWF, Conservation International, CARE, the Environmental Defense Fund (EDF), and Partners with Melanesians. The projects aimed at several different objectives, including biodiversity conservation and carbon storage, using several innovative methodologies. These initial efforts totaled $16 million, drawn from current-year funds in anticipation of the full $150 million in the next fiscal year budget. But few if any other wealthy countries went in with us on this effort. Skepticism? Free riding? Hard to say.

In late January 1993, the Clinton–Gore administration took over the government. The first week of the new administration, its new head of environmental policy, Katie McGinty, and I had lunch in my office to discuss FFI and related matters. Over the next few days, despite a pledge to jettison and rethink the entire international environmental policy portfolio of the Bush administration, the new Clinton–Gore administration, and Katie McGinty in particular, championed FFI. The new director of environmental matters at OMB had been a WWF economist, and he was sympathetic. At CEA, we remained committed to the idea. I urged choosing a new name, to make it the Clinton–Gore team's "own" initiative, but Katie liked the name Forests for the Future. In February 1993, President Clinton's first budget submission to Congress requested $50 million for FFI. That was not the

$150 million announced at Rio, but it was far better than losing the initiative altogether.

Still, it was lost. The Democrat-controlled 103rd Congress was on a mission to reduce the deficit, and a prime area for cuts was foreign aid. In mid-1993, the funding for FFI was dropped by the Senate, despite pleas from Katie McGinty and OMB.

CONCLUSIONS

Global forest conservation remains an urgent need and an institutional vacuum. Forests are a perennial third-stringer to climate and other sexier environmental issues. Perhaps this is a tragic result of the frustrating politics of collective action. Or perhaps our childhood fairy tales depicting forests as evil, such as Red Riding Hood, are so deeply ingrained that we cannot muster much enthusiasm for their protection.

But there may still be room for a policy entrepreneur, at CEA or elsewhere, to find a new avenue for mobilizing global forest conservation. I remain optimistic that we can one day create a global system of tradable forest conservation obligations, and hence a market-based system for "beneficiaries pay" financing of forest conservation.

A new global forest conservation treaty along these lines not only would try to solve a serious global problem, but also would herald the advent of a novel kind of global property rights. These global property rights would be "hybrid" in the sense that they were created by overarching administrative action (the treaty) instead of by local activities blessed by judicial action, the traditional route for the emergence of property rights.[46] And they also would be "hybrid" in a second sense: they would bridge the divide between formal legal property rights systems (such as tradable permits) and informal common property systems derived from shared norms (such as community management of a shared forest).[47] At the international level, formal legal property rights systems must of necessity arise from common property systems based on shared norms—that is, from the consensus of the community of nations. Thus, a treaty creating tradable forest conservation obligations would constitute a new form of property: "hybrid–hybrid" property created at the global level to internalize global public goods. That would truly be a Schumpeterian innovation worthy of the twenty-first century.

NOTES

1. See Stiglitz (1997, *109, 112*).

2. Ibid., *111, 113*.

3. Ibid., *112*.

4 From 1989 to 1991, I worked at the Environment Division of the Department of Justice. In January 1992, I accepted the offer to join CEA, effective July 1992. In the meantime, I joined the Office of Science and Technology Policy (OSTP), one of CEA's sister offices in the White House complex. The Rio Earth Summit was held in June 1992.

5. See FAO (1999, *1*). From 1990 to 1995, the rate of net loss was only slightly slower: 13 million hectares (0.7 percent) lost per year in poorer countries, 1.76 million hectares (0.1 percent) gained per year in wealthier countries, and a net loss of 11.26 million hectares (0.33 percent) per year worldwide (ibid). Total forest cover was 3.511 billion hectares in 1990 and 3.454 billion hectares in 1995 (ibid., *135*, Annex 3, Table 3). (The same data are reported in WRI 1998, *292*.) About 55 percent of the 1995 total forests were in poorer countries and 45 percent were in wealthier countries (ibid., *1*). In 1995, about 26.6 percent of the world's land area (13 billion hectares) was forested (ibid., *130*, Annex 3, Table 2). This was down from 27 percent in 1990.

6. WRI (1998, *187*) estimates that humans have cleared half of the forests on earth over the last 8,000 years; Yoon (1993, *C1*) reports that many forests currently thought to be "virgin" are actually regrowth after clearing by humans that occurred hundreds or thousands of years ago.

7. Warrick (1998, *A4*) reports on a survey of 400 biologists, two-thirds of whom agreed that a mass extinction is currently occurring and up to 20 percent of all living species could disappear within three decades.

8. This depends on local land users being able to reap the benefits of their investment in forest conservation over time. In countries where land users lack secure title to property, they may overconvert forests to nonforest land uses because they do not internalize the benefits of conservation that require more than a few years to reap, such as long-term soil fertility. See Hecht (1993); Deacon (1994).

9. See Coase (1960, *1*); Demsetz (1967, *547*).

10. See Ascher (1999); O'Toole (1988).

11. Ascher (1999) finds glimmers of hope in some case studies. Becker gives a theoretical case for such hope (1983, *371*). For further discussion of this thesis, see Wiener (1999b, *749, 757–58 & nn. 38–39, 792 & n. 179*).

12. A nice illustration of these expanding levels of forest conservation benefits, and their relation to the inadequate incentives for conservation, is provided in Stone (1995, *577, 580–88*).

13. In addition, poorer countries may find such a proposal unfair, given the history of forest conversion in wealthy countries, and intrusive on their national sovereignty.

14. For a more thorough analysis of why international environmental protection must be based on a "beneficiaries pay" approach rather than a "polluters pay"

approach, see Wiener (1999a, *677*, *735–55*). Other analyses taking this view include Stone (1995, *588–96*, *618–20*); Humphreys (1996, *163*); Sandler (1997, *95–97*); Swanson (1994).

15. Stockholm Declaration of the United Nations Conference on the Human Environment, Principle 21, 11 I.L.M. *1416, 1420.*

16. 27 U.S.T. 1087, 993 U.N.T.S. 243 (March 3, 1973; entered into force July 1, 1975).

17. See Swanson (1994).

18. 31 I.L.M. 849 (May 29, 1992; entered into force March 21, 1994).

19. 37 I.L.M. 22 (December 10, 1997; not yet entered into force).

20. Stewart and Wiener (2003, *87, 116*); Wiener 1999a, *757*, discusses perverse effects of the CDM on net emissions; ibid., *774–75*, concerns the CDM undermining the accession of developing countries to emissions caps; ibid., *794–95*, looks at the possible centralization of authority in the CDM.

21. 31 I.L.M. 818 (June 5, 1992; entered into force December 29, 1993).

22. See Sedjo (1992).

23. See WCF (1999).

24. See UNCSD (1997); ibid., para. 142, notes the absence of a comprehensive international legal instrument dealing with all types of forests; ibid., para. 143, reaffirms sovereign right of countries over their natural resources; ibid., para. 147, considers options including a new international legal instrument on all types of forests.

25. See UNCSD (1999).

26. See Humphreys (1996, *21, 136–45*); WCF (1999).

27. See Elwell (1996).

28. FAO (1999, *2–5*); Allan and Lanly (1991, *30*). See also IPCC (1990, *95*). On the government policies encouraging shifting cultivation as a cause of forest loss, see Ascher (1999); Hecht (1993); and Deacon (1994).

29. Allan and Lanly (1991, *30*).

30. Schneider 1996, *36*, reports a UN study finding that 9 out of 10 tropical trees are felled for agriculture or ranching; of the remaining 10 percent cut for the wood, only 4 percent (i.e., 0.4 percent of all trees felled) are taken for timber, the rest going to local fuelwood; and of the 0.4 percent felled for timber, less than a third (i.e., less than 0.14 percent) enter international trade. Indeed, as the Food and Agriculture Organization (FAO) defines its terms, logging in the tropics does not cause deforestation, because tropical logging is mostly selective (not clear-cuts) and by itself does not change the land use to nonforest. Allan and Lanly (1991, *29*). Still, according to IPCC (1990, *96*), selective logging in the tropics may damage 30 to 70 percent of the remaining trees, and the construction of logging roads may invite subsequent colonization and clearing.

31. FAO (1991).

32. The FCCC and Kyoto Protocol, by focusing only on the carbon storage service of forests, could similarly encourage plantation forestry (rapidly storing carbon) as a replacement for mature old-growth stands (not rapidly storing carbon), especially in the tropics, where clearing the old growth would not liberate as much

carbon from the soil as it would in temperate forests. The result could be an unintended depletion of biodiversity in the rush to store more carbon for global warming protection. This suggests another reason for the urgency of a forest conservation treaty that would embrace both carbon sequestration and biodiversity protection.

33. See Wiener (1999a, 757–60).

34. See ibid., 785–87.

35. As part of this obligation, Stockholm Principle 21 might be modified or rewritten in a forest convention to say, "States have the sovereign right to manage their forests in accordance with national developmental policies and priorities, the responsibility to support the provision by other states of shared environmental benefits from forest services, and the reciprocal responsibility to ensure that activities within their jurisdiction or control do not erode the value of forest services to the global community."

36. See Dudek et al. (1992) for surveying examples and cost savings.

37. See 42 U.S.C. 7651–7651o.

38. See Wiener (1999a, 726–27, 755–57), arguing that pure subsidies for international environmental protection may be perverse.

39. An allocation is necessary under any forest convention aimed at results. For example, a convention requiring all states to stop forest loss, or to adopt certain forest management principles, implicitly allocates all of the burden to forest-holding states. Here, the allocation would be more explicit but no more divisive—indeed, perhaps less so.

40. An interesting question is whether a system of tradable obligations to conserve a certain amount of forests would yield results equivalent to a system of tradable permits to clear forests. The latter could be allocated as a declining quantity of annual permits, accomplishing a phasedown of net forest clearing to some acceptable annual level. Poorer countries would have to receive extra permits as compensation for the cost of limiting their clearing; wealthier countries would receive fewer permits and could purchase some from poorer countries, thus generating a transnational "beneficiaries pay" financial flow. Further study is needed on whether the obligations approach would protect forests more effectively than the permits approach. I am grateful to Stewart Schwab for discussion of this question.

Other options for accomplishing the payment to forest holders might include a central fund, similar to the Global Environment Fund (GEF), in which the allocated obligation would be akin to a global tax going into the fund. The main drawback of this option is its dependence on a central bureaucracy, with few incentives to choose cost-effective projects, and much risk of market power. See Wiener (1999a, 794–96). A modified central fund might allow contributors to satisfy their obligations through direct bilateral assistance projects reported to the fund (similar to the Montreal Protocol fund); this would help force the central fund to compete to seek cost-effective projects. Still, a general system of decentralized bilateral arrangements is likely to be the most cost-effective.

41. On the need for widespread participation to prevent "leakage," the relocation of forest-converting activities to nonparticipating countries, see Wiener (1999a, 692–97 & n. 67).

42. See Weitzman (1974, 477).

43. For elaboration, see Wiener (1999a, *760–71*).

44. See Weitzman (1974).

45. See Interagency Task Force (1993, *1*).

46. See Rose (1998, *129, 163–66*); Stewart (1990, *91, 93*).

47. See Rose (1999).

CHAPTER SIX

ELECTRICITY RESTRUCTURING AND THE ENVIRONMENT

Stephen Polasky

$\mathbf{A}$t first glance, deregulation issues do not seem to be particularly relevant for environmental policy. Whether a particular industry should be deregulated typically revolves around questions about market organization and performance, not environmental issues. At the beginning of 1999, it also was not clear that deregulation of electrical generation was an important federal issue. State utility commissions are responsible for administering most of the regulations concerning electricity generation, and many states were already moving ahead with deregulation plans.

Despite surface appearances, however, important and difficult environmental and federal issues were raised by the prospect of deregulating electricity generation. In fact, some of the most contentious issues within the administration regarding deregulation had to do with environmental provisions, particularly the treatment of renewable power sources. In addition, federal issues arose because electrical power is transferred across state lines and because several federal agencies generate significant amounts of electricity. Federal legislation was needed to clarify how federal power agencies would fit in a deregulated electricity market.

Because of the rapid pace at which many states were moving on deregulation and the need to clarify the federal role, pressure was building for Congress to consider energy deregulation legislation. The Department of Energy (DOE) was charged with formulating the administration's position. Formally, the administration's proposal was described as electricity restructuring rather than deregulation, as there was little in the way of federal deregulation, but much in the way of changes in the structure of the federal role that would allow a smooth transition to deregulation of electricity generation by states. In 1998, the Clinton administration put together a proposal of electricity restructuring. Certain provisions of the proposal were subject to intense debate within the administration. This debate slowed release of the proposal to Congress, making it unlikely that it could pass legislation prior to adjournment. Congress failed to act on electricity restructuring in 1998.

During winter and spring 1999, the Clinton administration revised its proposal on electricity restructuring to cover several areas not adequately addressed the previous year, as well as to make changes reflecting new information. The process culminated in the public release of the administration's position at a news conference in April 1999. Speakers at the news conference included Secretary of Energy Bill Richardson, EPA administrator Carol Browner, and CEA chair Janet Yellen. Several members of Congress also were present, spanning the political spectrum from liberal Edward Markee (D–MA) to conservative Steve Largent (R–OK). Having these members present illustrated that electricity restructuring was an issue on which there could be bipartisan agreement.

THE DEREGULATION OF ELECTRICITY GENERATION

Since the time of Adam Smith, economists have extolled the virtues of the market. In the best of all possible worlds, and in introductory economics textbooks, the invisible hand of the market guides competitive producers to create what consumers want at the lowest possible cost. In other words, the market generates an efficient outcome. There is a presumption among many economists that it is best to leave production and consumption activities to the market with minimal government intervention, even when we are indeed in less than the best of all possible worlds.

There are, however, several exceptions to this general presumption of minimal intervention. Externalities, which are at the heart of the field of environmental economics, are important exceptions. Because a generator of an externality such as air or water pollution inflicts harm on others that the generator does not pay for, too much pollution will be generated in an unregulated market outcome. Environmental laws of some form are necessary to prevent this type of market failure.

Another important exception is monopoly power. Only under perfect competition do markets generate efficient outcomes. Monopolists curtail output to raise price, thereby generating an inefficient outcome. Antitrust laws are designed to prevent the exercise of such power.

In some cases, however, it is less expensive to have a single firm producing a good or service. For example, it is cheaper to have a single firm with a single set of electrical wires responsible for the distribution of electric power to end users than to have two or more firms with multiple sets of electric wires through the same area. In such cases of "natural monopoly," in which it is efficient to have only one producer, the preferred government response has been to regulate the natural monopoly rather than to apply antitrust laws to force competition. Electric utilities have long been thought to be classic examples of natural monopolies. State public utility commissions have regulated the generation, transmission, and distribution of electric power, along with some responsibility for regulation at the federal level.

Beginning in the 1970s, there was a reevaluation of economic regulation. Some industries that were being regulated were not natural monopolies. Economists argued that deregulation of such industries could lead to increased competition and improved performance. In the Carter and Reagan administrations, many industries were deregulated. These industries included airlines, interstate trucking, and long-distance telephone communication. Although critics of deregulation in these industries exist, the general consensus is that deregulation has improved performance. Overall, prices in these industries have fallen, sometimes dramatically, without declines in quality of service, loss of service to important segments of the population, or increased concerns over safety.

The electric power industry then came under the same scrutiny. Could all or part of the electric power industry be deregulated? The electric power industry can be thought of as being composed of three distinct segments: generation, transmission, and distribution. Transmission, moving large amounts of electricity at high voltage, and distribution, delivering electricity to end users, are both classic examples of natural monopolies. Generation, however, is not a natural monopoly. Although power generation plants tend to be large, these plants are not large relative to the size of the market. Numerous generation plants typically serve major metropolitan areas. Advances in transmission allow electricity to travel farther with less loss of power, thereby increasing the scope for competition among producers. Further, advances in power generation technology have lowered the cost of small-scale power production relative to large-scale production. The only segment of the electric power industry that is subject to deregulation and competition is generation. Electricity deregulation then is really deregulation of the generation of electricity with continued regulation of transmission and distribution.

Because transmission and distribution would still be regulated, some provisions had to be made to allow competing companies to be able to deliver electricity to consumers. To facilitate competition among generators, the Federal Energy Regulatory Commission (FERC) issued rules that required any interstate transmission line to allow any company to use the transmission and distribution lines at cost (FERC Order 888). Most states have similar rules covering intrastate transmission and distribution.

THE 1999 ELECTRICITY RESTRUCTURING PROPOSAL

With states moving ahead rapidly with the deregulation of electricity generation, most in Congress and the Clinton administration agreed there should be federal legislation on electricity restructuring. Since the administration had already put together a proposal in 1998, the 1999 version was supposed to be wrapped up quickly, with only minor details to work out. But perhaps because it was viewed as being a high-priority item on the congressional agenda, with a reasonable shot at legislation being enacted, various groups lobbied hard to get an administration proposal favorable to their interests, and agencies within the administration pushed hard for provisions as well.

Three issues in particular became the center of much internal administration debate: the renewable portfolio standard (RPS), the treatment of federal power generating agencies in a deregulated market, and the treatment of rural areas. The treatment of rural areas was a concern of the Rural Utilities Service of the Department of Agriculture, which works with rural electric cooperatives, many of whom feared competing in a deregulated market. The issues involved in the treatment of rural areas, though interesting, did not have important environmental aspects. The other two issues, RPS and the treatment of federal power generating agencies in a deregulated market, both had important environmental dimensions. The fights over these two issues are interesting in their own right, but they also illustrate some general points about the roles of economic analysis and of politics in environmental policy within the Clinton administration.

THE RENEWABLE PORTFOLIO STANDARD

The generation of electricity from renewable power sources, such as wind and solar, are more environmentally benign than generating electricity by other means. Coal-fired electricity generating plants emit large quantities of sulfur dioxide and nitrous oxides, which are precursors of acid rain and ozone, sometimes called smog. They also emit mercury. In addition, burning coal is a major source of carbon dioxide emissions, which are linked to global climate change. Burning oil or natural gas also contributes to local air pollution and global climate change through emissions of carbon dioxide. Nuclear power has risks associated with reactor accidents and disposal of radioactive wastes. With renewable power sources, there are no emissions from electricity generation (except from the burning of biomass and from production of the necessary equipment) and no waste disposal problems. Further, because the power source is renewable, valuable nonrenewable resources are not used up to produce electricity. Strong support existed within the Clinton administration for encouraging renewable power generation.

At the time of the discussion of the administration's electricity restructuring proposal, a substantial preexisting tax credit was available for electricity generation from renewable sources (1.5 cents per kilowatt hour). Given the negative externalities associated with competing sources of generation, preferential treatment of renewable power generation is justified. Strictly speaking, economic theory suggests that the best policy would be to tax sources generating negative externalities rather than to subsidize sources that do not create such externalities. Under the tax approach, the price of electricity will reflect the full marginal social cost of electricity generation. Under the subsidy approach, the price will be less than the full marginal social cost, leading to overuse of electricity. Early on, the Clinton administration suffered a major political defeat in proposing to increase taxes on energy. Despite concerns over air pollution, the difficulty the

United States would face in meeting its carbon dioxide emissions targets under the 1997 Kyoto Protocol, and concerns about energy security, proposals for taxes on energy sources were off the table. Advocating higher energy taxes was the surest way to be uninvited to future meetings.

The electricity restructuring proposal was seen by parts of the administration as another avenue by which to promote renewable electric power generation. The RPS would require that each company generate a certain percentage of power from renewable sources after a phase-in period. Renewable power sources under the RPS would include solar power, wind power, and power from burning biomass. Because of environmental controversies surrounding hydroelectric dams, the RPS excluded hydropower. In 1998, after a heated debate, with EPA and the Council of Environmental Quality (CEQ) on one side and the Treasury Department, the Office of Management and Budget, and CEA on the other, the RPS was set at 5.5 percent. In 1999, EPA and CEQ argued that the RPS should be increased far above 5.5 percent, perhaps as high as 10 percent; the economic agencies contended that the RPS was already too high and should not be raised further. High-ranking officials from across the administration debated in numerous meetings about how high to set the RPS. They finally reached an agreement to set the RPS at 7.5 percent. Currently, less than one-half of 1 percent of electricity is generated by sources that fit under the RPS.

The intense fight over the RPS was really much ado about nothing. The Republican leadership in Congress had made it clear that they would ignore the administration's proposal regarding the RPS. How high to set the RPS in the proposal was then really a symbolic rather than a substantive policy issue. One thing I learned in D.C., however, was how seriously people took fights with symbolic meaning, even when no substance was involved. In modern politics, it is more important to look good than to be good. In an age of instant media and short attention spans, policy that can be summarized by a pithy sound bite triumphs over better, but harder to explain, policy alternatives.

Even ignoring what Congress would do, the administration's electricity restructuring proposal was written in such a way that increases in the RPS would have minimal or no real impact. This surprising result occurs because of the combination of two special features contained in the administration's proposal. First, a safety value provision in the RPS allowed a company to buy credits from DOE for 1.5 cents per kilowatt-hour when it could not generate enough renewable power to meet the RPS. According to estimates by DOE, companies would be highly unlikely to meet the RPS if it was set above 6 percent. Increases above 6 percent would result in companies purchasing more credits from DOE rather than generating more electricity from renewable sources. Second, a different part of the administration proposal specified that companies would be assessed a surcharge on electricity generation, which would go into a Public Benefits Fund. The total amount of money that could go into the fund each year was capped. Money raised from sales of RPS credits also would count against the cap. So raising more money via credits would mean raising less money via the surcharge. Not only would an increase in the RPS not result in more generation from renewable sources, but there also would be no change in the aggregate revenue that the government would raise from utilities.

Although during most of the year that I spent at CEA I was struck by how different the world of Washington politics was from academia, in some ways the fight over the RPS reminded me of fights within the university. It has been said that fights among academics are especially bitter because so little is at stake.

FEDERAL POWER GENERATION AGENCIES

Beginning in the New Deal era, the federal government became involved on a large scale in the generation of electricity. This involvement began with hydroelectric dams but grew to include fossil fuel and nuclear electricity generating facilities. Originally the dams were built to provide electricity to regions with little or no existing electricity, to aid economic development in those regions, as well as to provide irrigation and flood control. Several independent federal agencies were set up to oversee the operations of the electricity generating facilities. Four of these agencies have significant electrical generation capacity. The Tennessee Valley Authority (TVA) is the largest electricity generator in the country, producing about 50 percent more power than the largest investor-owned company. The Bonneville Power Authority (BPA) is also of large size. The other two, the Western Public Power Authority and the South West Power Authority, are of moderate size.

Deregulation of electricity generation raises questions of competition between the federal power generation agencies and private investor-owned utilities. The federal agencies worry that under deregulation, because they provide services other than electricity and must serve all of the public, including high-cost rural consumers, they might not be able to compete with investor-owned utilities, especially for the most lucrative industrial and commercial customers. On the other hand, investor-owned utilities point out that federal agencies do not pay taxes, have the full faith and credit of the U.S. Treasury to back their loans, and have other advantages not available to private companies. Investor-owned utilities in the Southeast were quite concerned about unequal competition between themselves and TVA. Some details about how the federal power generation agencies should be treated in a deregulated environment were ironed out, but larger issues of whether there was a level playing field or fair competition were not on the table for discussion. At one point in discussions, I was told that the "T" in TVA stood for "Tennessee," and that Vice President Gore's office would handle sensitive issues related to these questions.

One important issue with environmental aspects arose in the discussion of BPA under deregulation. For most of its history, BPA has supplied low-cost hydroelectricity from its series of dams on the Columbia and Snake Rivers. A number of recent factors have threatened to raise BPA's costs of generating electricity. Several runs of salmon in the Columbia River basin have been listed under the Endangered Species Act. Although there are other causes as well, most observers think the dams on the Columbia and Snake Rivers have played a substantial role in the decline of salmon in the basin. BPA currently spends approximately $400 million per year on species protection programs. Whether to remove four dams on the lower Snake River also was debated. Removing these dams would increase BPA expenses further. In addition to salmon-related expenses, BPA acquired large debts from a bad investment decision in the failed Washington Public Power Supply System (WPPSS). A series of low-water years in combination with high species protection costs and debt repayment could lead to a situation in which BPA would not be able to cover its costs in a deregulated market.

Economists might answer that if BPA could not compete, then it should not be in the business of power generation. After all, one of the points in favor of market competition is that it drives production costs as low as possible and pushes the inefficient out of the market. The hydroelectric dams in the Pacific Northwest really are a low-cost source of electricity, however. If BPA cannot cover its costs, this will be largely because of the WPPSS debt. If BPA should not be able to repay the WPPSS debt, either the bondholders or the U.S. Treasury (taxpayers) will pick up the tab, the latter being much more likely.

BPA, however, did not like the economists' answer, arguing that it should be allowed to put a surcharge on the transmission of electricity should circumstances arise in which it could not cover generation costs (and the WPPSS debt). BPA controls much of the transmission capacity in the Pacific Northwest. Competitors to BPA would have to use this transmission capacity to supply customers in the region. A transmission surcharge, therefore, is anticompetitive and runs directly counter to the deregulation effort. CEA, along with the Departments of Treasury and Justice, expressed reservations. In spite of the economic arguments against the transmission surcharge, BPA managed to prevail and get the surcharge provision into the administration's electricity restructuring proposal. BPA gained support from the environmental and resource agencies, EPA, CEQ, and the Department of Interior, because it was able to convince other agencies that support for salmon recovery might be jeopardized if BPA was not financially solvent. In this way, the transmission surcharge was portrayed as being proenvironment. BPA also gained support from the region's political leaders, who worried about the financial health of BPA even though the surcharge would potentially raise electricity prices for the region's consumers. A potential future surcharge, however, was viewed as more palatable than raising current prices (which are well below market rates) to create a reserve fund, even though a reserve fund is a more sensible alternative on economic grounds than the transmission surcharge.

CONCLUSIONS

Environmental policy affects many issues, even those that at first glance appear not to be environmental issues. Electricity restructuring is ostensibly about rewriting federal rules governing the regulation of electric utilities to allow states to proceed with deregulating electricity generation. Debates over the renewable portfolio standard and the BPA transmission surcharge, however, illustrate that environmental issues are important in debates over electricity restructuring. These debates also illustrate that a policy that appears to be good for the environment, even if it will not really do anything to improve environmental conditions, may be politically beneficial. Such policies may be harmful in other dimensions, however, as the example of the transmission surcharge shows.

CEA has the unenviable task of arguing for economic efficiency in the political arena, where efficiency considerations are not a high priority. On specific decisions, economic analysis often plays only a minor role in environmental policy. Economic arguments are used more where convenient to buttress a decision that has already been made on political grounds than to inform the decision-making process. On broader policy directions, however, the imprint of economists and economic ideas is much more evident. Tradable emissions permits have fundamentally altered regulation under the Clean Air Act, and deregulation itself is an idea that economists have pushed for several decades.

To many observers, recent events have cast doubt on the wisdom of pursuing deregulation of electricity generation. The actual record to date of electricity deregulation is anything but impressive. California, a pioneer in deregulation, experienced skyrocketing electricity prices and rolling blackouts in 2000 and 2001. One of the state's largest investor-owned companies, Pacific Gas & Electric, filed for bankruptcy in April 2001. The state of California was saddled with large costs incurred from efforts to resolve the electricity crisis. On August 14, 2003, the power went out on 50 million people in the Midwest and Northeast. By most estimates, the blackout was the worst ever recorded. Critics contend that deregulation was largely to blame for California's woes and factored into creating conditions that made a large-scale blackout more likely. These problems have led many to call on Congress and state governments to stop electricity deregulation and provide increased regulation aimed at making the electricity system more reliable and prices less volatile.

The problems plaguing the California electricity market were partly due to bad luck and poor planning, but bad policy decisions also played a large role. Demand for electricity was growing rapidly, while at the same time little new power generation or transmission capacity was added to the system. On top of the already tight conditions, limited hydroelectric power supply plus a large fraction of the generation capacity in many facilities in California being shut for repairs were enough to trigger blackouts in early 2001. Policy mistakes played a key role in creating and intensifying the crisis. Wholesale prices were deregulated but retail prices remained regulated until March 2002. The California utilities complained that they had to pay close to $1 per kilowatt-hour on the wholesale market while they were limited to charging 6.5 cents on the retail market. Uncertainties over future policy directions gave little incentive to investment in new generating or transmission capacity.

Although I am generally supportive of efforts to deregulate electricity generation, the developments in California and elsewhere show that deregulation done badly is worse than not doing it at all. Natural monopoly in transmission and distribution, concerns over reliability of the electricity grid, and environmental issues mean that the correct approach to electricity is not to "let the market work" with no government intervention. Done well, however, deregulation of generation can lead to more efficient production and lower prices for electricity in the long run. No compelling economic logic suggests that regulation is necessary for electricity generation.

The challenge in designing electricity restructuring is to facilitate competition in electricity generation within a comprehensive policy approach to the electricity industry. Regulatory oversight is needed to ensure fair and open access for all competitors to transmission and distribution lines. There is a role for policy to provide the proper incentives to maintain the capacity and reliability of the interconnected electricity grid. Adequate antitrust oversight is necessary to prevent firms from gaining market power in the market for power. Environmental regulations need to provide incentives to reduce the negative effects associated with pollution emissions from electricity generation. Public policy may be needed on assistance to the poor should price spikes cause undue hardship. Deregulation of generation by itself does not address the wider set of issues. This is why a comprehensive electricity restructuring proposal, rather than simple deregulation, is necessary. These issues should be addressed to ensure that deregulation of electricity generation leads to an improvement rather than a decline in performance.

The dreadful record of the past few years is not an indictment of deregulation itself. Rather, the recent record illustrates the importance of systematically thinking about policy design for the electricity industry. An important part of this thinking must be about the incentives (or lack thereof) that policy provides to both producers and consumers of electricity. Economists at CEA can play a vital role by pointing out flaws in various policy alternatives as well as providing suggested solutions that should lead to better performance.

CHAPTER SEVEN

DO ALL THE RESOURCE PROBLEMS IN THE WEST BEGIN IN THE EAST? REVISITED

Jason F. Shogren

In his 1992 book, *Where the Bluebird Sings to the Lemonade Springs*, Wallace Stegner notes that "one of the things Westerners should ponder, but generally do not, is their relation to and attitude toward the federal presence. The bureaus administering all the empty space that gives Westerners much of their outdoor pleasure and many of their special privileges and a lot of their pride and self-image are frequently resented, resisted, or manipulated by those who benefit economically from them but would like to benefit more, and are generally taken for granted by the general public."

Stegner raised a question in my mind that nagged throughout my 1997 stint as the senior economist for environmental and natural resource policy at the Council of Economic Advisers (CEA) in the White House, and continues to do so today: do all the resource problems in the West begin in the East? By the West, I mean the interior rural West, including the Rocky Mountain states and the eastern parts of Washington, Oregon, and California; by the East, I mean Washington, D.C. Although Riebsame (1997) stresses that the area considered the West keeps moving around in time and space, the interior West stretches roughly from the foothills of the Rocky Mountains to the crests of the Sierra Nevada and Cascade Range.

As I witnessed the nature of federal policymaking inside the Clinton–Gore administration, I wondered how well-intentioned advocates for social regulations affected locals trying to get by based on their traditional ways. The effort expended by advocates troubled me in light of the administration's strategy of pursuing a centralized environmental agenda that is more unyielding toward the West in order to secure support on the coasts. Imposing more federal will on the West through tougher administrative actions gathers more bicoastal green votes. This struggle continues today, although the Bush–Cheney administration sees the West less as a colony for recreation and more in its traditional role as a domestic energy colony. Western senators continue to remove some centralized management strategies developed in the East by attaching prodevelopment riders to different appropriations bills. Which party is in the White House seems not to matter—both development and conservation of resources in the relatively unpopulated West are still run by those outside the area. Regardless of the party in power, I continue to believe that economics can help make good resource policy better and bad policy go away.

Back in 1997 at CEA, I began to understand better the role and power of economics in the message we delivered and our role as messengers. And though my tone now is less pitiless, the lessons hold up. The depths of western discontent regarding federal regulations became concrete while I was at CEA (also see Shogren 1998a,b). Events such as the Sagebrush Rebellion in the early 1980s, in which many westerners demanded more input into local public land decisions, made more sense to me (see Cawley 1993). I better understood their frustration with how local resource management and control were affected by decisions made in Washington, D.C. (For different perspectives toward stewardship, see Anderson and Leal 1997; Berry 1992; Ciriacy-Wantrup 1952; and Robbins 1994.)

Rules debated and codified in Washington, D.C., made me wonder how and why economic reasoning, as taught in the textbooks, had such a seemingly small influence on thinking about how these new rules could affect the country, the West in particular. On reflection, however, I now see that economic reasoning was used every day; political tradeoffs over the costs and benefits between administration actions and inaction over environmental protection and other policy goals were simply made way beyond my pay grade at CEA.

Were all the pundits who push for states rights and local control correct? Was this the century of the state, in which the federal leviathan ruling the West had left us worse off than if we had just taken care of ourselves? Why are private expenditures to comply with social regulation exceeding $200 billion and rising—large enough that Robert Litan of the Brookings Institute (quoted in Passell 1998) points out that "federal regulation now costs consumers more than total discretionary government spending costs taxpayers"?

Living in Wyoming for nearly a decade has revealed many of the contradictions that define the West. In 1998, the *Economist* magazine dubbed Wyoming the most conservative socialist state in the Union. My friend Tom Crocker calls this split personality "socialism for me, free enterprise for you." This schizophrenic nature of the West is real: a clamoring for more control over setting the standards for regulation, but still wanting to preserve the hard-won resource subsidies that keep the traditional livelihoods, such as ranching and timber, intact. My family is no different; we want our independence, yet we appreciated the federal help in moving from the backwoods to the blackboard. It is a mix of well-subsidized classical liberals who promote individualism and still ask for the federal government to help pay for the time-honored lifestyles.

Before serving at CEA, I was apolitical and had little appreciation of the split personality of the West. While common sense waxed and waned in Washington, D.C., unproductive ideas would eventually collapse under the deadweight losses they created. The people we elected and the agencies we allowed to run our public affairs were doing a good enough job for me, and I did not need to worry about whether the West should be more autonomous. The ins and outs of economic theory were more interesting than Washington.

After my time at CEA, however, I understand that some people see economic orthodoxy as not simply confining, but downright prehistoric. Degaudio (1997) is quite animated about this point: "[T]he people have little interest in the ideas of the professors of economics. . . . [T]he people have little patience for academic incrementalism and less stake in the excessively cute laws concocted to make federal agencies do paperwork." When people think they are correct about a particular policy direction, they do not need costs or benefits to confirm their vision about the future. Forcing them to confront the potential opportunity costs seems to them like a step backward in time. And it became clear that my lapses in communicating key economic concepts left the floor open for arguments that federal control of local issues just made more sense.

For instance, the Environmental Protection Agency (EPA) estimated the retrospective costs and benefits of the Clean Air Act Amendments (CAAA) between 1970 and 1990. Section 812 requires EPA to assess how the Clean Air Act has affected "public health, economy, and environment of the United States" and to report these findings to Congress (U.S. EPA 1997a). EPA's self-evaluation found that federal regulatory actions created substantial net benefits from lower-criteria air pollutants. Thus the EPA inspector general's 1998 pronouncement of widespread failures by state and local officials to police the nation's clean air was a bit unexpected (e.g., Cushman 1998). The mixed messages send the signal that although EPA is doing well, it could do even better if more environmental affairs were handled at the federal rather than the state or local level.

It is news like this that makes me think economics is merely a speed bump on the road of bad ideas. I wonder whether the arguments leveled at economics and economists are accurate, and why economists' counterarguments are less than persuasive. Just how limiting is our orthodoxy, and why do we have a hard time changing the minds of critics who consider economics "not so much dismal as half-witted" (attributed to Sir Crispin Tickell in the *Economist* 1997)? Perhaps many economists are overoptimistic about the power of economics; this might suggest that more awareness of what outsiders are saying to us would be helpful. If we want to hone our skill as advocates for more protection at less costs, allowing people in the West to create their own luck, we need to revisit and reemphasize the sophisticated informality that can help make that happen.

IN DEFENSE OF ECONOMICS IN RESOURCE POLICY

In policy debates in Washington, D.C., economists who ask about opportunity costs are tagged as "lemon suckers." Ideologues and moralists underappreciate economists who point out the reality of scarcity. These critics make assertions about the limits of the economist's choice of tools to cast doubt on the message, and about the economist's toolbox to cast doubt on the messenger. Unanswered assertions stick, regardless of their veracity. Following are eight allegations made, explicitly or implicitly, in resource policy debates, along with my responses. It would have been better to have provided these responses sooner, but quick retorts from me were not always

forthcoming, given my background in the classic Nordic school of nonde-
bate. I hope that these responses to common complaints will help better
prepare those economists interested in becoming less apolitical.

1. *Economics always overestimates the costs of federal regulation. It is likely
that the estimated effects on the U.S. economy of the Kyoto climate change treaty
are simply too high.* The evidence does not support this assertion. Squitieri
(1998) and Harrington et al. (1999) have compiled the actual evidence,
and they find that costs are on the low side relative to predicted levels. Pre-
dictions straddle actual costs for asbestos, coke ovens, and vinyl chloride
regulation. Numerous unpredicted changes in technology and the economy lowered the cost of CFCs, cotton dust, and SO_2 control. The National
Ambient Air Quality Standards were estimated to be achieved by 1977, but
these standards still have not been attained in 75 areas with a total of 75
million people. Modelers predicted that the control costs for sulfur dioxide
could be $1,500 per ton, when today a ton actually costs $100. Reasons
exist for this gap—unanticipated technology breakthroughs, railroad deregulation, permits given out for free. Costs are not systematically biased
upward. This implies that the cost estimates to implement the Kyoto treaty
could be low. Kyoto calls for industrial nations to reduce their carbon emissions by about 7 percent below 1990 levels by the period 2008–2012, without the inclusion of the developing nations, whose emissions are projected
to surpass those of today's industrial nations within the next three or four
decades. Imagine turning a battleship on a dime with a third of the crew on
board, and you get a good feel for the potential costs of the Kyoto climate
change treaty.

2. *Economics downplays risks to children's health and welfare. Look at the current federal air pollution standards for ozone and particulate matter. President
Clinton supported the tighter standards, even though the costs might have
exceeded the benefits, because he wanted kids to be healthy. We are for kids.
Whose side are you on?* Economics is for kids. The goal is to create policies so
that children have food in their bellies, clothing on their backs, roofs over
their heads. Economics promotes these ends through promoting rules to
help in the creation and storage of wealth. At issue is how the insights from
economics about the direct and indirect links between wealth and health
can be more clearly communicated without seeming callous. Children
deserve special attention, and that does not necessarily imply more federal
control over state and local resource policies. Pointing out that the conspic-
uous neglect of behavioral rules can inflate the costs of regulation at no
gain in kids' health does not sell politically if the route is too indirect. The

appearance of helping kids, even at the price of taking away aid through more regulation, will still be more appealing to policymakers than the roundabout route to wealth through more trade. This is not to say that economic theory cannot improve on how we model and estimate risks to children. Economics would benefit from more age heterogeneity in our models, since kids rarely have economic standing even when they have legal standing.

Economics also has to address the severe limitations in data on household behavior. Behrman (1997) identifies several limits to the data needed for intrahousehold resource allocation decisions. First, individual human capital endowments are not observed in most socioeconomic data sets, making it nearly impossible to directly estimate how different endowments affect behavior. Rarely observed are the intrahousehold allocation and distribution of transfers and investments such as self-protection, non–labor market time, nonschool time, food, exercise, and the quantity and quality of caregiver–child contact. Also, available data usually covers a short window of behavior, whereas many questions of interest span an entire lifetime, and data do not link children with adult siblings who live outside the household. Many key outcomes influenced by household allocation and distribution decisions are unobservable, inestimable, or both, such as the rate of return on human capital investments. Finally, measurement error in most observable variables is a challenge to assess and correct.

3. *Economics oversells individualism and exaggerates the ability of locals to solve their own problems in the West. If locals could solve their own problems, why were federal acts such as the Endangered Species Act (ESA) and the National Environmental Policy Act (NEPA) enacted in the 1970s?* Over the last decade, the costs imposed by such supralocal intervention into private property have triggered devolution of power back to more local levels. The ESA, for instance, is one of the most extreme forms of federal intervention, the quintessential prohibitive policy. In its shadow, people have turned to cooperative conflict-resolution mechanisms that involve mainly local stakeholders. These mechanisms have flourished despite traditional environmentalists who remain convinced of the need for federal intervention because of their skepticism that locals will neglect local habitat. New field experiments in collaborative decision making have emerged in which local people are engaged in the communication, education, and cooperation involved in the ESA and NEPA processes (see, e.g., Brown and Shogren 1998; Gould et al. 1997; Kite et al. 1998). This place-based collaboration encourages local stakeholders to resolve local environmental disputes

through bargaining and negotiation, rather than through litigation or by federal mandate. Devolution of environmental regulation allows communities to develop and enforce consensus management plans that adopt specific place-based strategies that complement federal legislation. In principle, collaboration offers the potential for an efficient construction of local environmental protection. This trend will likely continue. Economics has not exaggerated the role of the local individual in helping define and implement rules everyone can live with to deal with resource issues in the West.

Collaboration groups number in the hundreds, ranging from informal grassroots gatherings to government-mandated advisory councils. The western vision of collaboration is the driving force behind Enlibra, the Western Governors Association's doctrine for environmental management in the region. The governors want less remote control and more local control over western resources. Enlibra outlines their push for strong local leadership to balance development and conservation goals, and to resolve environmental conflicts.

The first Enlibra principle is "national standards, neighborhood solutions—assign responsibilities at the right level." Locals understand local conditions. Rather than come up with unimaginative bureaucratic responses, the federal government should help local people and policymakers develop their own plans to achieve binding targets and to provide accountability. The second principle is "collaboration, not polarization—use collaborative processes to break down barriers and find solutions." The western governors believe that community-based collaboration can help produce creative solutions with political momentum. Together these principles support local leadership and collaborative efforts to help landowners and others enhance the environment and achieve economic productivity.

But the western governors take this doctrine a few steps further. Enlibra does not hold fast to one tool as the means for effective and accountable local control of natural resources. Consensus works in some cases but not in others. The governors recognize that a variety of tools in combination with collaboration can be used to improve western environmental and community well-being. "Markets before mandates—pursue economic incentives whenever appropriate" is the relevant Enlibra principle. In some cases, collaboration might be better organized with an auction block than at a bargaining table.

4. *Economists are too pessimistic about how people can adapt their preferences to new technologies to solve resource policy questions such as climate change, in*

which engineering studies suggest that from 20 to 25 percent of existing emissions of the greenhouses gases that cause climate change could be eliminated at no additional costs. In effect, economists take a creationist view toward preferences. The open question is the degree to which people will do the right thing without having to be prodded by a change in relative prices. A better-informed U.S. populace could achieve the Kyoto targets by changing their minds about the risks of climate change. Economists are skeptical of these arguments because they usually see people as people, with similar preferences across the ages and around the world. When the evidence says that people prefer conventional appliances given current prices, economists believe it. They also believe there is a fundamental distinction between people's preferences and a true market failure. The market does not fail when people focus on immediate outlays even when given future cost-savings estimates, or when they are wary about claimed energy savings that might go unrealized. Economists see that the most effective way to curb excessive energy consumption is to raise its price to reflect the harmful effects on the environment of burning fossil fuels. Sure, economics can benefit from understanding the origin of preferences, but for the most part, preferences are taken as stable given the exchange institution in which they operate.

But whether one thinks price changes are needed depends in part on what Sowell (1987) calls the "choice of vision." People with an unconstrained vision believe we all have a vast, untapped morality buried within waiting to emerge with the right direction. Thus solutions such as the Kyoto protocol are primary, the tradeoffs involved secondary. As Sowell puts it, "Every closer approximation to the ideal should be preferred. Costs are regrettable, but by no means decisive." People with a constrained vision, however, weigh ideals against the costs of achieving them. Real incentives are needed to get people to take on some goals behind environmental protection. The uncertainties underlying many environmental questions leave latitude so that whether a person chooses to believe that the benefits justify the costs rests on his or her choice of vision.

5. *Economics downplays the discontinuities and disequilibriums of nature.* There is cogency here in that resource policy discussions eventually reach the point at which the economist is asked whether he or she has accounted for the likelihood that a change in the ecosystem will be discontinuous as a result of a catastrophe. Most economists acknowledge but do not always incorporate into their models the potential for discontinuous risks, such as a sudden shift in the Gulf Stream because of climate change or an unraveling of the web of life because of the loss of some keystone species. But does

this mean that society is on the cusp of catastrophe? Not necessarily; the doomsayers have a terrible track record. As revealed in numerous risk perception studies, people commonly overestimate the chance that they will suffer from a low-probability, high-severity event, such as a nuclear power accident (see, e.g., Viscusi 1992). When the outcome is potentially very bad, people inflate the chance that the outcome will be realized. This tendency transfers to policymakers as well. They overestimate the risk themselves, and then multiply this impact by playing off the same tendency in the general public. For example, resource policies toward the storage of nuclear waste in the West can conjure up images of a fortified storage facility containing sanitized, airtight receptacles or an abandoned dump site teeming with rusty, leaking vats of toxic material. These images induce vivid perceptions, both of which can persist in a community, causing considerable disagreement about how to regulate the risk. Policymakers must act as the arbiters who help reconcile the perceptions, not inflame the differences.

6. *Economists have a built-in bias toward the status quo business world and act as mercenaries for the corporate bottom line.* Economists can blame themselves for the perception that they are good at telling some people's stories but not others. Economists have allowed the general media to portray economics as synonymous with financial and commercial concerns. When a person hears about the local promotion of economic development, it sounds as though economists deal strictly with financial and commercial development. But economists worry as much about Elm Street as about Wall Street and need to better communicate the broader perspective of welfare economics to the general public. Both development and conservation have value; the question is how much of each and where and when. For example, cost–benefit analysis can justify either development or conservation of resources, depending on the preferences of people in society. If the benefits of development are too low relative to the costs—in other words, too many endangered and threatened species are sacrificed—economics would promote conservation over development. Numerous economic reasons exist for preservation: some species and habitats provide useful goods and services, others are valued aesthetically, and even seemingly low-value species are linked to high-value species through ecosystem interactions. Economists need to help people better appreciate that economics goes beyond timber markets and does try to include the nonmarket values that make our lives fascinating.

7. *Economists who promote cost–benefit analysis fail the elementary political test of persuasion, because such efforts increase the scale of political controversy by*

entrenching conflict by venturing to put values on losses inconsolable by money. Most economists agree with Hayek's forceful point that the market is the most powerful tool ever discovered by humans to integrate efficiently the diffuse set of knowledge that exists throughout the economy. Market prices reflect both the laws of man and the laws of nature. As with language, no one invented the market, but it has emerged as the promoter of more rational choices on how to allocate scarce resources. Cost–benefit analysis simply takes this rationale one step further by collecting and organizing diffuse knowledge into a common framework for goods and services left unpriced by the market. If one agrees that the market works as a knowledge magnet that facilitates rational choices and cooperation, it is self-defeating to argue against cost–benefit analysis, because it works under the same principle.

8. *Economists need to realize that science drives policy but morality sets policy. Economics is most useful when it sticks to estimating the costs to implement the policy selected.* I believe that this statement is wrong. Economic circumstances affect whether a species is endangered and whether it should be listed, because human adaptation to both economic and biological conditions affects the odds of species survival. Most people expect, for instance, that more preserved habitat implies greater odds of species survival. Human communities are characterized by key economic parameters such as wealth and the relative prices of land. Communities with greater wealth or lower relative land prices can better afford to preserve more habitat. Ignoring wealth, land prices, and other economic parameters when estimating the odds of species survival is to omit relevant variables.

But economists themselves have turned over too much authority to the natural sciences in resource policy decisions. Wilson (1998) claims that economics is largely irrelevant to resource problems because economists avoid the complexities of the "foundation sciences" such as psychology and biology. But the circle of environmental knowledge is also broken when ecological and biological models ignore the effects that the economic system has on the natural system. Natural science alone cannot accurately answer whether we confront an environmental problem. Economics contributes to the answers because relative market prices and wealth play vital roles in establishing the threshold that indicates whether an environmental issue is a real problem. The so-called foundation sciences need to connect mind to matter by including economic parameters in their core frameworks. Economists have a responsibility to correct this omission in debates of environmental policy—to help define the environmental thresholds of human and ecosystem health that underpin policy, rather than settle for a secondary role. Economists need to move their feet further under the table where resource policy and research are defined. They need to clarify that they accept as a necessary fiction rather than as literal truth the economist's orthodoxy of rational maximization and equilibrium to help noneconomists understand that economists know the power and limits of the tools of their trade.

COLLABORATION

I have learned the difference between a scientist and a mad scientist. A scientist experiments on others; a mad scientist experiments on himself. After returning to Wyoming from D.C., I ran for and was elected to our county school board and was appointed to the state's Environmental Quality Council and the Western Governors Association's advisory board for Enlibra. I say this not for sympathy, but to illustrate my newfound commitment to both making decisions that matter and better understanding the pressures at work when these decisions have to be made. I appreciate the sense of seriousness that drives many economists. Our role goes beyond accountants of the axiomatic. I worry less about whether, say, the value of recreational fishing per day trip is $3.54 or $4.12; rather, I worry about whether economists should let the moralists and ideologues make all the big decisions. And we do take ourselves out of the real-time debates by trying only to impress each other. For as hard as it is to convince anonymous reviewers of one's ideas, it is harder to dejargon and communicate these ideas clearly to people who have to make decisions. But we do have a responsibility to point out that new rules remain undesirable unless the benefits exceed the costs.

Do I still believe the West suffers because economists are misunderstood in policy debates the East? No. The West is schizophrenic, and no economic theory is going to change that. Snow (1997) calls the West the region of overworked adjectives—Wild, Old, New, Next, Real, and so forth. People in the West have two visions: the "pull-yourself-up-by your-boot-straps" path, in which the lone cowboy self-image is a reality, and the "cow-boy-with-complex-preferences-with-his-hand-out" path, in which the West grudgingly accepts intervention for goodies. For the "lone cowboy," economics says go ahead and blame the East for ignoring sound arguments against bad rules, push for devolution, but do away with subsidies that define the landscape. This path presupposes that the willingness of west-erners to accept responsibility and accountability exceeds that of the absen-tee landowners, private and public, who currently dominate the western landscape. An open-space future for the West then depends on our own willingness and ability to pay for the empty space we cherish. For the "com-plex cowboy," however, economics says that the West has only itself to blame for more federal intervention, and that westerners must continue to work with the East to set goals, review which subsidies stay and which go, and accept a watered-down version of devolution. The West still must come to terms with what it is and what it wants.

Dan Kemmis, director for the Center of the Rocky Mountain West at the University of Montana, has such a vision. He believes the time is right for the West to govern the West. Similar to the semiautonomous relationship that Scotland and Wales have with the United Kingdom, Kemmis sees the rural interior West recapturing its democratic roots through authentic self-determination. He calls for a "progressive, bipartisan, regional solidarity among western governors and within the Senate . . . for a West-wide com-pact to transfer responsibility for public lands to western institutions" (2001, 231). This ambitious vision appeals to many westerners. And per-suasive economic reasoning can provide solid information to help guide whether semiautonomy makes sense for the West. Economists can con-tinue to demonstrate and communicate that behavior matters more to resource policy than most people think. With additional empirical tradeoff analysis that sharpens the economic viewpoint, we can increase the costs to policymakers who neglect or downplay the importance of behavior in resource policy. Policymakers can benefit from clean evidence that reveals how resource policy could be less expensive when accounting for relevant economic behavior.

CONCLUSIONS

Economists have made significant inroads into resource policy at the federal level. Successes such as the push to codify cost–benefit regulatory reform in Congress and the endorsement of tradable carbon emissions permits in Kyoto demonstrate that good ideas endure over the long run. But economics has its own problems that deserve more attention. Economic historian Mark Blaug attributes the problems of economists to being too easily distracted from reality to escapist puzzles and second-order technicalities (see Blaug

1998; also see Krugman 1998 for a response that promotes the idea of sophisticated informality).

We economists suspend the craft of diplomacy and inflate our status in the discourse of public policy by giving too little credibility to the existence of political "wise men." The scorched-earth approach driven by economic correctness comes across as callous and arrogant, and it can be counterproductive in resource policy settings. And though we do not have to be wallflowers, more emphasis on the aggressive, sophisticated informality as used by Schultze and Keynes and Friedman to advocate efficiency seems most worthwhile.

This all makes sense and we should do more of it. How? One way to start is to accept our responsibility to inform others, rather than expect others to automatically understand us. Taking one's ideas that have survived the professional review process, recasting them in everyday language, and publishing them in mainstream outlets where ordinary people might read them is valuable (see, e.g., Holt and Bielema 1996). Some economists do this already; more should probably try, including me. Krugman (1997) is the current role model for this practice. Another step I have taken is to have my graduate students write "memos to the president" that translate what they are doing into plain English and explain why it matters for society. If the current short-term incentives in academia do not exist to take these steps, we should do it anyway. We have to care enough about our ideas to get them out into the light.

For 5,000 years, the best humans could do was to increase our life expectancy by five years. About 200 years ago, however, something changed, and since then Western culture has witnessed a 30-year increase in how long we might live. Was it is just a coincidence that around the same time, The Wealth of Nations was published? I do not think so. For two centuries, economists from Adam Smith to Hayek to Friedman have argued for the market as the best way to organize the diffuse set of information and to direct motivations in society. The market is a process of discovery, a creator of wealth, and more wealth creates more health. Many people would be willing to give economists some of the credit for at least a few of those 30 extra years. We economists can continue to do good things if we keep working to give people a chance to invest in themselves so that they can create their own luck. Behavior matters, and economists who can explain why in simple and well-timed language can help make resource policy better, in both the East and the West.

NOTE

This chapter revisits views that the author originally expressed in the article, "Do All the Resource Problems in the West Begin in the East?" which was the keynote address at the 1998 meeting of the Western Agricultural Economics Association and was published in the *Journal of Agricultural and Resource Economics* in December 1998. The author thanks Tom Crocker, Sean Fox, Sally Kane, Sherrill Shaffer, Ray Squitieri, the reviewers for helpful comments, and Bruce Weber for identifying the Stegner quotation.

CHAPTER EIGHT

THE GOOD NEWS AND THE BAD NEWS FROM WASHINGTON

Ray Squitieri

This chapter is about good news and bad news. The good news is that economists are increasingly occupying positions of influence, not only in Washington, but in other national capitals as well. Former treasury secretary Larry Summers is not the first academic economist of high reputation to hold a cabinet post. George P. Shultz left his position as dean of the School of Business at the University of Chicago to serve in four cabinet posts, including secretary of treasury and state. But Larry Summers and his colleague Joseph Stiglitz, who served as chairman of the Council of Economic Advisers (CEA) a few years ago, are among the most distinguished academic economists of their generation. My agency has been able to recruit intellectual horsepower and economic expertise at the subcabinet level at a magnitude unmatched during my 10 years in Washington. The same thing is occurring in other countries. In Russia, former prime minister Yevgeny Primakov has a Ph.D. in economics. In Mexico, former president Ernesto Zedillo has a Ph.D. from MIT and only 10 years ago was an economist at the Central Bank. The level of economic literacy at the top of governments is the highest it has been at any time in my career.

The bad news is that all this economic talent has not led to an outbreak of rationality in government decision making, and particularly not on environmental issues.

Stiglitz (1998) says that when he came to work in the White House, he often felt that he had arrived in a different world. He had expected this world to have its own language, as every culture creates its own specialized jargon. What he did not expect was that this world would have its own system of logic. Important decisions were made on the basis of a few anecdotes, standards of evidence were low, and much of the evidence offered was simply irrelevant.

Economists who have spent time in Washington quickly recognize the aptness of Stiglitz's observations. It would be nice to report that the failings he cites appear mainly at the staff level, and that quality improves as the issue moves to higher levels. But often the reverse holds true. The staff work may be good and the debate at the staff level respectable, but then the quality of the debate declines as the issue moves up the decision-making ladder.

POLICY ANALYSIS IN ACTION (OR OUT OF ACTION)

Here are a few examples from my own experience.

- I keep a file that I call "Not Clear on the Concept." One item is a memo from an assistant attorney general to her peers in other agencies proposing to extend the community right-to-know law to greenhouse gases. This law is designed to protect a community from harmful emissions that may come from factories in the neighborhood. But because atmospheric gases mix together quickly, and because CO_2 has a long residence time in the atmosphere, greenhouse gases are a worldwide, not a local, issue. The extra molecule of CO_2 emitted in Laramie, Wyoming, has exactly the same effect on the atmosphere above Laramie, Washington, Beijing, or Timbuktu. So if a right-to-know law requires companies to report their emissions of greenhouse gases to people who live near the plant, then it should also require letters to the rest of the six billion people on the planet.
- In amending the Clean Air Act of 1990, Congress told the Environmental Protection Agency (EPA) to conduct a retrospective cost–benefit

assessment of the act from 1970 to 1990. Researchers who have tried to measure the costs and benefits of different sections of the act have arrived at different conclusions; some sections appear to be net winners, others net losers. One might expect EPA's assessment to also tell a mixed story. Instead, EPA's retrospective assessment calculated large net benefits for every section of the act, with total 1970–1990 net benefits of $23 trillion. On hearing this, a fresh-out-of-college CEA junior staffer pointed out that $23 trillion is roughly the value of the entire U.S. capital stock, and that if EPA's result were true, it would mean that Americans should be indifferent between having 1970-level air quality and the existing capital stock and having 1990-level air quality and no capital stock.[1] Paul Portney, president of the nonpartisan think tank Resources for the Future, labeled EPA's result preposterous.

- The federal government has subsidized renewable fuels, mostly ethanol, for two decades. Originally explained as necessary in order to allow ethanol to compete with price-controlled petroleum fuels, ethanol subsidies persisted long after the end of oil price controls in 1981. The result has been a continuing costly subsidy to a fuel that is twice as expensive as its competitor, gasoline, and three times as expensive as conventional fuels in electricity generation. The environmental benefits are questionable as well; EPA staff has consistently opposed expansion of the program, but the subsidy has continued. Internal debates during both the first Bush and early Clinton administrations pitted an antiethanol coalition of the Treasury Department, CEA, Office of Management and Budget (OMB), and often EPA and the Department of Energy (DOE) against only the Agriculture Department, which offered weak and unconvincing arguments. Both times, the antiethanol coalition believed it held all the cards. Both times, it lost.

I do not mean to suggest that all would be well if we economists were put in charge. It is the job of economists to provide objective analysis based on measurable factors. It the job of politicians to make decisions that take into account unmeasurable factors as well. The good news is that economists are increasingly occupying positions of influence. The other good news is that the economists are not in charge.

Any economist who spends time in Washington will see a lot of missed opportunities: win–win opportunities to improve public policies that are not adopted. Why is it so difficult to implement these potential Pareto improvements, asks Stiglitz (1998), changes that could, in the end, make the whole society better off? Why is it so difficult to put in place policies that would improve everyone's welfare, even after the parties are aware of the advantages of the proposal? But the more complicated situation in Washington requires some institutional details.

Several years ago, CEA and the Congressional Office of Technology Assessment put together a proposal to promote biomass fuels, which would have eliminated the current ethanol subsidy. In the view of the economists and other technical people involved, this proposal was in every way superior to the current program. The only losers would have been the ethanol producers, and they could have been compensated with the savings generated by the new proposal. The proposal was defeated, however, when the ethanol lobby mobilized its mighty battalions of lawyers and political supporters.

Superfund is EPA's program governing the cleanup of old toxic waste sites. Most economists who have studied Superfund have concluded that it has imposed costs out of proportion to the benefits to public health. A comprehensive study conducted by Hamilton and Viscusi (1999) and funded by EPA supports this view. A good portion of the outlays have gone to litigation expenses; according to a 1992 study by the Rand Corporation, only 12 percent of insurance company outlays actually went to cleaning up sites. Moreover, Superfund imposes costs on people and firms that have only the remotest responsibility for any damage that might have been caused. One of our deputy assistant secretaries had to recuse himself from Superfund discussions because his grandfather had, for two years in the 1940s, owned a tanning plant in Florida, and he could have inherited money at his grandfather's death. Decades later, the site was declared a Superfund site, and all previous owners of the site were named potentially responsible parties in the case, whether or not they had violated any laws on the books at the time.

If any environmental program offers the potential for a Pareto improvement, it is Superfund. Nonetheless, attempts at fundamental reform have consistently stalled. The 103rd Congress (1993–1994) came close to reauthorizing Superfund. The administration had sent up a bill, with bipartisan support, that would have modestly improved the program. But the law did not get to the president's desk, mostly because insurance companies could not agree among themselves on cleanup and liability provisions. In the 104th Congress, the new Republican chairmen of the key authorizing committees made a determined effort to eliminate retroactive liability; this failed when no one could find a way to pay for this step. In the 105th Congress, both Republicans and Democrats evidently believed that, despite wide areas of agreement, they had more to lose by advancing a reform proposal than by sticking to the status quo. Even in 2003, when Republicans commonly portrayed as caring little for the environment controlled the administration, the House, and by a slim majority, the Senate, there was no active discussion of meaningful reform of Superfund.

Olson (1971) argues that because a concentrated interest group has lower costs of organizing itself than a large, diffuse group, the small group is likely to prevail. I think of this as a theorem: *all else being equal, the small group wins.* That is, a zero-sum game pitting a small group against a larger group will be won by the small group. The mission of CEA (and a few other offices sprinkled around the government, including my office at Treasury) is to redress this built-in imbalance, or at least to point out whose thumb is on the scale. The "small group wins" theorem explains why the ethanol producers (or the dairy lobby, or the orange growers, or the trial lawyers) usually win in a zero-sum game, but many cases are positive-sum games where more is involved.

Stiglitz (1998) points out that many proposals actually offer potential Pareto improvements, or near-Pareto improvements, as he calls them, offering benefits to a large, diffuse group that exceed the losses to a small group. The large group may offer some part of its potential gain to compensate the small group for its loss, but the small group may have no way to enforce the agreed-on side payment.

The designers of Washington's new metro system considered alternatives to the expensive (and temperamental) elevators built for the use of the disabled, including on-demand free or low-cost jitney or taxi service for disabled patrons. The disabled lobby rejected this proposal because, first, it would have made the subsidy transparent, possibly undermining its political support, and second, the metro authorities could not make a credible commitment that the subsidies would continue. The same kind of inability to make a binding commitment stalled the proposal to eliminate ethanol subsidies, as the recipients of current subsidies naturally worry about exposing gains to public scrutiny. For several decades, U.S. airlines received cash subsidies to carry airmail; the airlines loved the subsidies but hated their transparency.

Other reasons for the failure to achieve potential Pareto improvements, according to Stiglitz, are coalition formation and bargaining. The Coase theorem tells us that if transaction costs are small, the market will reach an efficient allocation of resources. But with imperfect information, meaning that transaction costs are not small, the outcome may not be efficient. This is the case of Superfund, where the parties do not have good information about the other parties' real preferences, willingness to pay to achieve desired outcomes, or acceptable tradeoffs. When the parties in a public policy debate rely on the negotiating techniques of an oriental bazaar, it is a safe bet that they are using inefficient signals.

Stiglitz also cites destructive competition, meaning the conversion of a positive-sum game into a zero-sum game by the intervention of political considerations. Supporting a compromise proposal may allow Congressman A to help his constituents, but because Congressman B might get the credit for solving the problem, Congressman A withholds his support. This occurs frequently in Washington.

These ideas do much to explain the apparently irrational outcomes one often sees in environmental policy debates. They help explain how agents acting in their own best interest repeatedly leave money on the table by rejecting potential Pareto improvements.

ANALYTICAL THINKING AND INTUITION

Some questions have puzzled me about the debate itself. Answers may lie in the relationship between epistemology—how we know what we know—and public discourse, particularly as it applies to environmental topics.

When analysts—including economists—get involved in public policy, the underlying questions generally concern the actual state of the world: Do fine particulates in the air cause people to drop dead, and if so, how many, and with what loss of life-years? Does TRIS, a flame retardant once used in children's clothing, cause cancer? Is the earth getting warmer, and if so, is the Kyoto Protocol a sensible remedy?

We say loosely that these are questions about truth. But what is this thing we call truth? And what do we have when we find it?

Truth often refers to knowledge arising from systematic rational thought and disputation. But truth also may refer to knowledge arising from an inner sense. This kind of truth is subjective, personal, and direct. A landscape architect once mentioned to me that a certain tree was a Norwegian maple. I asked him how he knew. He thought for a moment, then answered, "That's like asking me how I know that someone is my brother. I could say that I know he is this tall and weighs this much and has a birthmark here on his neck, but the real answer is that I just know."

When analysts talk about truth, they are referring to a kind of under-standing arrived at within the framework they accept. The rules and the process are important: exposition of hypothesis; presentation of evidence and argument, including empirical testing according to established rules; acceptance of contrary evidence and argument; conclusion. Analysts accept the requirement to go through these steps before claiming that something has been established to be true. They may or may not believe that what they are saying is true at some deeper level. Voltaire reminded us that his-tory is a fable agreed upon. All of us involved in public debate have agreed to the collective "fables" of our own disciplines (economics, ecology, epi-demiology) and the "fables" of our society, such as "all men are created equal." The formal rules and the agreed-on "fables" serve the important purpose of permitting a debate to occur.

Not surprisingly, there is uneasy coexistence between subjective personal truth and the objective social process required for the truth that emerges from analysis.

This brings me to the mystic and the scholastic. The word *mystic* is likely to conjure images of crystal balls and faux oriental costumes. But an older sense of the word refers to the acquisition of knowledge by the direct expe-rience of God. Joan of Arc heard voices telling her to drive the English out of France. Like most saints, Joan was a mystic. Mystical knowledge comes not from deductive reasoning, rational disputation, or empirical observa-tion, but from an inner sense. Joan's inner sense took the form of voices. She was also a field commander of great effectiveness; she could not have been just twiddling her thumbs all day and waiting for God to tell her when to attack the English right flank. So being a mystic does not preclude other ways of gaining knowledge.

In contrast, Joan's thirteenth-century predecessor Thomas Aquinas was a scholastic. Aquinas attempted to understand the things that mattered—God, man, and the universe—by applying logical rules that rational men could all agree on. He had no need for the special knowledge of the mystic. His *Summa Theologica* attempted to explain all of Christian doctrine in the framework of Aristotelian logic. Aquinas's work is still recognized as one of the great achievements of Western philosophy, even though most people no longer buy his line of argument.

Aquinas's system was tailor-made for the hierarchic and bureaucratic Catholic Church of the High Middle Ages. More than 700 years after his death, he remains the dominant philosopher of the Catholic Church. His system provided an orderly, logical, and systematic way to understand the world. It provided a way to examine questions of Catholic doctrine and practice. It provided a way to examine new claims. And it provided a framework for debate, allowing Catholics with differing views to discuss these views, debate their differences, and arrive at a conclusion.

For Aquinas and other scholastics, mystical knowledge was irrelevant, and mystics were an irritation. If mystics got their knowledge directly from God, then their claims were not subject to the formal scholastic system. Nor did the mystics need priests, bishops, or the rest of the Church apparatus. Thus, the Catholic Church as a political institution has always had trouble dealing with mystics and mystical knowledge (although long after they are dead and no longer a threat, mystics might be declared saints).

With our tradition of personal liberty, many Americans are inclined by temperament to side with the mystics. Thoreau, that most American of American philosophers, was a mystic, not a scholastic. He would not tolerate any large bureaucracy telling him how to think and behave, whether Harvard University, from which he dropped out, or the state of Massachusetts, from which he also dropped out, in a sense, when he refused to pay his tax to support the Mexican War.

But there are advantages to the scholastic approach, particularly in a modern urban and pluralistic society, when people with very different inner experiences, and very different ways of seeing the world, must debate public issues and reach agreement.

Aquinas's scholastic system, elaborate and fussy though it may be, is a good model for public debate in our era. Most Westerners no longer share Aquinas's view of the universe, but we have adopted a scholastic model as a way to conduct our public debates.

Relating this to environmental policy, a productive public debate on any issue requires a set of ground rules. We generally use ground rules that I will loosely call the secular scientific view: secular because it excludes knowledge available exclusively from revealed religion; scientific because it uses the conventions of scientific inquiry and discussion. This is our modern equivalent of the scholastic rules set out by Aquinas: a set of ground rules for inquiry and discussion, consistent with the prevailing worldview. Thus, our modern scholastic rules demand more empirical testing than either Aquinas or Aristotle required.

In our public role of providing objective analysis according the guidelines of our discipline, economists are scholastics. We arrive at knowledge (at least we claim to have arrived there) by certain formal methods of inquiry, and we present and debate our results with others who are bound by the same methods. At home, however, we may be mystics. In our private decisions—whom to marry, what career to pursue, how to handle a difficult teenager or a life-threatening medical emergency—we rely on intuition, hunches, and other knowledge that would not be admissible in our public role.

Sometimes intuition functions as a sort of superanalysis. Good chess players use intuition because the analytical mind cannot account for the thousands of possible positions that could result from each move. Presidents of companies and presidents of nations use intuition because it allows them to incorporate a greater variety of information in their decisions than they could with purely analytical thinking. Economists and scientists use intuition to steer their research, beginning with an inner sense that something is important, and then acting on that sense.

Intuition as superanalysis is useful for political decisions, which must account for many factors simultaneously, many of them difficult to measure. Secretary Robert Rubin once overruled me and my office on an issue we cared deeply about. I felt stung and believed that he would have come around to our view if only he had known the facts. Much later, I realized that he probably did know the facts, or at least most of them, and that his decision reflected tradeoffs we had not been privy to.

As a general rule, intuition and analytical thinking do not mix well in policymaking. At the staff level, debates take place according to the scholastic rules. At the political level, however, the game is different, and the scholastic rules may no longer apply.

At a meeting on global climate change, an assistant treasury secretary insisted on having estimates of the costs of the various options under consideration. The key White House political official responded by saying that the group must first decide on a policy (she had a particular one in mind), and only then figure out how to pay for it. Another meeting on another environmental issue featured a debate about the costs and benefits of various options. A different White House political official stopped this discussion by insisting that the benefits of her proposal were so important that the costs did not matter.[2]

In each of these examples, the debate had been proceeding according to the established scholastic rules but was cut short by a powerful official who attempted to impose her preferences on the group, either innocently, following some intuition about the truth, or cynically, understanding that the analysis was exposing flaws in her case. In either case, the appeal to intuitive knowledge short-circuited the analytical debate and prevented a full slate of options from reaching the president. To the analyst, it also seemed like a raw violation of the process.

The weakness of the scholastic approach is that the scholastics can make themselves irrelevant to the debate by focusing more and more on things about which people care less and less. This was the fate of the medieval scholastics, who carried on increasingly intricate debates about arcane matters of theological doctrine as the foundations of the debate were eroding beneath them; we still refer derisively to earnest debates about how many angels could dance on the head of a pin.

In our own culture, we occasionally see a radical alteration in the accepted model of the role of government. Kuhn (1962) describes a sudden radical alteration in the dominant scientific model as a paradigm shift; this term seems appropriate here, too.

The most radical such shift in U.S. history concerned the issue of slavery. In 1861, abolitionists made up a small minority in the North. When the Civil War began, no important Northern politician held out abolition as the main aim of the war. As the war ground relentlessly on, however, and astounding losses mounted on both sides, that began to change. In the North, the idea began to take hold that this mass slaughter could be justified only by a moral purpose bigger than merely preserving the Union; the abolition of slavery came to be seen as this great moral purpose. By 1865, public opinion in the North would no longer even consider a Union with slavery. Slavery, which just four years before had seemed like a distasteful regional quirk, had become anathema.

A less radical but still dramatic shift occurred during the first two terms of the Roosevelt administration. From the founding of the republic until 1933, the dominant model of government included only a few functions: defending the country's borders from external threats, providing a justice system to maintain internal order, providing civil courts to enforce contracts, and not much else. In his first inaugural address, Jefferson set out the model of a government whose major function was to prevent citizens from coercing each other: a "wise and frugal government, which shall restrain men from injuring one another, which shall leave them otherwise free to regulate their own pursuits of industry and improvement." Until 1933, the government's size reflected this view: from 1789 to 1933, government spending at all levels averaged around 5 percent of national income and, except during major wars, never exceeded 12 percent.

The idea of a more expansive government role began to take hold in Germany and England at the end of the nineteenth century, but this

attracted little support in the United States except among some intellectu-
als. By 1933, federal regulation had expanded to include a few sectors of
the economy, but few Americans expected that government would, for
example, transfer income to the unemployed or the indigent.

Roosevelt's first two terms changed those expectations. By 1941, government spending at all levels stood at 25 percent of national income, on its
way to more than 35 percent, where it remains today. Americans began to
expect the federal government to transfer income to the unemployed, the
indigent, and importantly, those who were merely old. Individuals, once
seen as responsible for their own fate, were now seen as being tossed about
by forces they could not control. In the turmoil, the government could no
longer be simply an umpire, but rather needed to be a participant.

OPPORTUNITIES FOR REFORM

Scholastic analysis serves several useful functions in policy discussions.
First, it generates valuable insights. Once introduced into the debate,
these insights tend to persist until they are addressed. Scholastic-style
analysis of competition and the effects of regulation formed the groundwork for the deregulation of airlines, telecommunications, financial services, and other industries in the late 1970s. More recently, scholastic-
style analyses such as those of the cost of Superfund or of the likely
effects of the Kyoto Protocol on climate change have fundamentally influenced recent policy debates.

Second, the scholastic method allows a forum for thoughtful debate,
because it has accepted rules. This is particularly important in environmental policy, which has long been sharply polarized on many issues, and
where thoughtful debate is often absent.

Scholastic analysis can be undermined in several ways, however. It can
be truncated by decisions imposed from outside the scholastic ground
rules, as in the examples above. Or it can be hijacked by political
appointees instructing their subordinates to make the analysis produce a
certain result. This gaming of the system undermines the credibility of all
scholastic-style analysis and causes a loss of faith in the ability of scholastics to inform the process. The scholastic process has weaknesses, to be
sure, but it is still the only game in town—the only set of rules available by
which to judge the merits of alternative policies. Discarding its insights
does nothing to make policymaking more rational. As often seen in recent

years, neglecting scholastic analysis can result in nonsensical regulations and can lead to divisive and costly court battles.

Naturally, the scholastic approach has its limits. It cannot identify the "best" regulatory policy (in economists' terms, "best" depends on the weights of the different factors in the utility function, and these weights are determined subjectively).[3] Scholastic analysis can be a part—but only a part—of political deliberations.

Certainly, we have better tools for scholastic-style analysis than we had 20 years ago: better data, faster computers, and better-trained graduates in economics and statistics. But have better tools led to more informed decisions? I would say no.

One reason is that the age of television has made scholastic-style analysis more difficult. Rational analysis has much to contribute to a print-based debate but very little to contribute to an image-based debate. Scholastic-style analysis cannot determine the truth or falsity of an image and is too slow to handle the flood of images washing over the citizen of the modern polis.[4]

In Europe in the early Middle Ages, a thing was true or false because the pope or emperor said so; this was the epistemology of pedigree. The scholastics of the late Middle Ages urged the epistemology of logic, emphasizing the value of human reason in deciding whether a claim was true or false. The scientific revolution of the sixteenth and seventeenth centuries advanced this view, and the Enlightenment advanced it still further, allowing even challenges to the authority of the Bible. The spread of printing facilitated the spread of scholastic-style inquiry, as it greatly increased the number of people with access to the arguments.

But the age of television has changed all that. In the print-based world of 1787, the authors of the Federalist papers could assume their claims would be debated on the basis of logic and evidence. Political debates today, however, are conducted in images, about which logic and evidence have little to say. Possessing neither the time nor the knowledge to independently assess the thousands of images we see each day, and the claims they imply, we rely instead on our opinion of whoever is putting them forward. Do they come from a trusted source, be it a certain newspaper, TV anchorman, or lobbying group? If so, we are likely to believe the claim. Thus, we revert to the epistemology of pedigree.

Public policy debates have moved gradually away from the language of logical discourse and toward the language of personal experience, with reliance on the endorsement of trusted sources. The internal administration debate over EPA's proposed National Ambient Air Quality Standards in 1997 featured abundant data, statistical analysis, risk measurements, and the other apparatus of scholastic analysis. The economic agencies were convinced that they had won the debate. In the end, this did not matter, as EPA moved the debate into the media, understanding that it had better images— children with asthma—and that a good image trumps good analysis.

CONCLUSIONS

Despite all this, there are some grounds for optimism. Within the federal government, several scholastic enclaves already enjoy protection from the vagaries of the political winds. These include the Energy Information Administration (EIA), within the Department of Energy; the Bureau of Labor Statistics (BLS), within the Department of Labor; and the Congressional Budget Office (CBO). (The Congressional Office of Technology Assessment, now defunct, also enjoyed this protection.) This protection gives the work of these offices considerable credibility. In contrast, less credibility attaches to analyses from the program offices of most agencies, which do their work at the pleasure of the executive suite. More independent offices along the lines of EIA, BLS, and CBO would improve the quality and utility of scholastic analysis in the policy debates. Also needed is better oversight of significant new regulations.[5]

Agencies have published a variety of official guidelines about when and how to conduct risk assessment and economic analysis. OMB and EPA have both issued such guidelines. Indeed the Bush administration in 2001 reaffirmed the economic analysis guidelines issued by OMB in 1996; and in fall 2003, it issued new guidelines reiterating the bulk of the recommendations in the earlier documents. These guidelines help protect the role for scholastic-style analysis by giving agencies that adhere to them and the scholastic rules they represent a chance to brag. The guidelines also give outsiders a chance to raise procedural complaints in those instances where agencies have deviated from their own guidelines and used exclusively mystic insights to justify policy choices. Finally, they reduce the risk that mystics will hijack scholastic inquiry by instructing subordinates to produce certain results. Analysis used to produce such politically motivated results presumably deviates from agencies' own guidelines. Politicians seeking to avoid embarrassment over such deviations may end up acting in ways that preserve both scholastic and mystic insights in policy decisions.

NOTES

The ideas expressed here are solely the responsibility of the author and do not necessarily reflect the views of the U.S. Treasury. The author thanks Keith Cole, Larry Goulder, Randall Lutter, Ed Murphy, Jay Shogren, and Peter Squitieri for helpful comments.

1. More precisely, Americans should be indifferent between having lived with 1970-level air quality trends extrapolated to 1990, and our current capital stock, and the actual level of air quality as it existed between 1970 and 1990, with no capital stock.

2. A Dilbert cartoon captures the analyst's frustrations well. The boss passes Dilbert in the hall and informs him that a management meeting has just made big changes to his project. Dilbert asks whether any of the technical staff had been asked to attend the meeting. "Oh no," says the boss. "We find things go more smoothly when no one with any actual knowledge is there."

3. For more on what this sort of analysis can and cannot do, see Arrow et al. (1996).

4. I am indebted to Keith Cole for this argument; as he notes, Postman (1986) made the basic point.

5. Although politicians from both parties say they favor better regulatory review, political forces push the other way, with the result that very little serious review ever takes place. New regulations received only the most cursory review under the Clinton administration. When the Bush administration took office in 2001, it promised to get serious about regulatory review, appointing John Graham of Harvard, a well-known expert on risk analysis, to head OMB's Office of Information and Regulatory Analysis. After a strong first year, however, the commitment to serious review seems to have waned, at least as measured by the number of rules returned to the agencies for further analysis or modification. Similarly, Congress established a regulatory analysis project within the General Accounting Office in 2000, but then never funded the office.

REFERENCES

Aldy, J.E., R. Baron, and L. Tubiana. 2003a. Addressing Cost: The Political Economy of Climate Change. In *Beyond Kyoto: Advancing the International Effort against Climate Change.* Arlington, VA: Pew Center on Global Climate Change, 85–110.

Aldy, J.E., S. Barrett, and R.N. Stavins. 2003b. 13+1: A Comparison of Global Climate Change Policy Architectures. Discussion paper 03-26. http://www.rff.org/rff/Documents/RFF-DP-03-26.pdf (accessed August 2003).

Aldy, J.E., P.R. Orszag, and J.E. Stiglitz. 2001. Climate Change: An Agenda for Global Collective Action. Paper presented at the Pew Center for Global Climate Change Workshop on the Timing of Climate Change Policies. October 11–12, 2001, Washington, DC.

Allan, T., and J.P. Lanly. 1991. Overview of Status and Trends of the World's Forests. In *Technical Workshop to Explore Options for Global Forestry Management, Bangkok, April 1991, Proceedings,* edited by David Howlett and Caroline Sargent. London: International Institute for Environment and Development.

Alliance to Save Energy, American Council for an Energy-Efficient Economy, Natural Resources Defense Council, Tellus Institute, and Union of Concerned Scientists. 1997. *Energy Innovations: A Prosperous Path to a Clean Environment.* Boston: Tellus Institute. June.

Ames, Bruce, and Lois Swirsky Gold. 1996. The Causes and Prevention of Cancer: Gaining Perspectives on the Management of Risk. In *Risks, Costs, and Lives Saved: Getting Better Results from Regulation*, edited by Robert W. Hahn. Washington, DC: Oxford University Press and AEI Press.

Anderson, T., and D. Leal. 1997. Rekindling the Privatization Fires: Political Lands Revisited. In *Breaking the Environmental Policy Gridlock*, edited by T. Anderson. Stanford, CA: Hoover Institute Press, 53–81.

Arrow, Kenneth, et al. 1996. *Benefit–Cost Analysis in Environmental Health and Safety Regulation*. Washington, DC: American Enterprise Institute, the Annapolis Center, and Resources for the Future.

Ascher, William. 1999. *Why Governments Waste Natural Resources*. Baltimore: Johns Hopkins University Press.

Barrett, Scott. 1994. Self-Enforcing International Environmental Agreements. *Oxford Economic Papers* 46: 878–94.

———. 1998. Political Economy of the Kyoto Protocol. *Oxford Review of Economic Policy* 14(4): 20–39.

Becker, Gary S. 1983. A Theory of Competition among Pressure Groups for Political Influence. *Quarterly Journal of Economics* 98: 371–400.

Behrman, J. 1997. Intrahousehold Distribution and the Family. In *Handbook of Population and Family Economics*, Volume 1A, edited by M. Rosenzweig and O. Stark. Amsterdam: Elsevier, 125–87.

Bernstein, P.M., W.D. Montgomery, T.F. Rutherford, and G. Yang. 1999. Effects of Restrictions on International Permit Trading: The MS-MRT Model. *Energy Journal: Special Issue on the Kyoto Protocol*, 221–56.

Berry, W. 1992. *Sex, Economy, Freedom, and Community*. New York: Pantheon.

Blackman, Allen. 2001. The Economics of Climate-Friendly Technology Diffusion in Developing Countries. In *Climate Change Economics and Policy*, edited by Michael A. Toman. Washington, DC: Resources for the Future.

Blaug, M. 1998. Disturbing Currents in Modern Economics. *Challenge* 41 (May–June): 11–34.

Bodansky, D. 2003. Climate Commitments: Assessing the Options. In *Beyond Kyoto: Advancing the International Effort against Climate Change*. Arlington, VA: Pew Center on Global Climate Change, 37–59.

Braine, Bruce. 2003. Sold! *Electric Perspectives*, 20–30.

Brown, G., and J. Shogren. 1998. Economics of the Endangered Species Act. *Journal of Economic Perspectives* 12(Summer): 3–20.

Browner, Carol. 1999. *Comments on the U.S. Court of Appeals Ruling on Ozone and Particulate Matter*. Conference at AEI-Brookings Joint Center for Regulatory Studies. May 27, 1999, Washington, DC.

Bruce, J.P., H. Lee, and E.F. Haites (eds.). 1996. *Climate Change 1995: Economic and Social Dimensions of Climate Change*. Second Assessment Report, Working Group

III, Intergovernmental Panel on Climate Change. Cambridge, UK: Cambridge University Press.

Burtraw, Dallas, and Erin Mansur. 1999. The Environmental Effects of SO_2 Trading and Banking. *Environmental Science and Technology* 33(20): 3489–94.

Burtraw, Dallas, Karen Palmer, Ranjit Bharvirkar, and Anthony Paul. 2001. *The Effect of Allowance Allocation on the Cost of Carbon Emission Trading.* Washington, DC: Resources for the Future.

Cawley, R.M. 1993. *Federal Land, Western Anger: The Sagebrush Rebellion and Environmental Politics.* Manhattan, KS: University of Kansas Press.

Cazorla, Marina V., and Michael A. Toman. 2001. International Equity and Climate Change Policy. In *Climate Change Economics and Policy,* edited by Michael A. Toman. Washington, DC: Resources for the Future.

CEC (California Energy Commission). 2002. *Renewable Energy Program, 2002 Biennial Report.* Sacramento, CA: California Energy Commission.

Ciriacy-Wantrup, S. 1952. *Resource Conservation: Economics and Policies.* Berkeley, CA: University of California Press.

Climate Action Network. 1997. *ECO: Climate Negotiations Newsletter* 18(6). http://www.climatenetwork.org/eco/c3_6_tactics.html (accessed December 6, 1997).

Cline, William R. 1992. *The Economics of Global Warming.* Washington, DC: Institute for International Economics.

Clinton Administration. 1998. *The Kyoto Protocol and the President's Policies to Address Climate Change: Administration Economic Analysis.* Washington, DC: White House.

Clinton, William J., and Al Gore. 1993. *Climate Change Action Plan.* Washington, DC: Executive Office of the President.

Coase, Ronald. 1960. The Problem of Social Cost. *Journal of Law and Economics* 3: 1–44.

Cohen, Richard E. 1992. *Washington at Work: Back Rooms and Clean Air.* New York: MacMillan Publishing Co.

Cooper, R. 1998. Toward a Real Treaty on Global Warming. *Foreign Affairs* 77(2): 66–79.

CEA (Council of Economic Advisers). 1998. *The Kyoto Protocol and the President's Policies to Address Climate Change: Administration Economic Analysis.* Washington, DC: Executive Office of the President.

———. 2000. *Economic Report of the President.* Submitted to the Congress, February 2000. Washington, DC: Government Printing Office.

Coy, Carol, Pang Mueller, Danny Luong, Susan Tsai, Don Nguyen, and Fortune Chen. 2001. *White Paper on Stabilization of NO_x RTC Prices.* Diamond Bar, CA: South Coast Air Quality Management District.

Cushman, J. 1998. EPA and States Found to Be Lax on Pollution Law. *New York Times,* June 7.

Daniels, Jr., Mitch. 2001. *Improving Regulatory Impact Analyses*. M-01-23. Office of Management and Budget. http://www.whitehouse.gov/omb/memoranda/m01-23.html (accessed June 19, 2001).

Deacon, Robert. 1994. Deforestation and the Rule of Law in a Cross-Section of Countries. *Land Economics* 70: 414–30.

Degaudio, E. 1997. Academia Usurped. *Regulation* (Summer): 6.

Demsetz, Harold. 1967. Toward a Theory of Property Rights. *American Economics Review* 57: 347–59.

DRI (Standard & Poor). 1998. *The Impact of Meeting the Kyoto Protocol on Energy Markets and the Economy*. Prepared for the United Mine Workers of America. June.

Dudek, Daniel J., Richard B. Stewart, and Jonathan B. Wiener. 1992. Environmental Policy for Eastern Europe: Technology-Based versus Market-Based Approaches. *Columbia Journal of Environmental Law* 17: 1–52.

Easterbrook, Gregg. 2001. Climate Change: How W. Can Save Himself on Global Warming. *New Republic*, July 23.

Economist. 1997. Plenty of Gloom. December 20.

———. 1998. The Wyoming Paradox. July 18.

Edmonds, J.A., H.M. Pitcher, D. Barns, R. Baron, and M.A. Wise. 1992. Modeling Future Greenhouse Gas Emissions: The Second Generation Model Description. In *Modelling Global Change*, edited by Lawrence R. Klein and Fu-chen Lo. Tokyo, Japan: United Nations University Press, 295–340.

EIA (Energy Information Administration). 2000a. *Analysis of the Climate Change Technology Initiative: Fiscal Year 2001*. Washington, DC: Department of Energy.

———. 2000b. *International Energy Outlook*. Washington, DC: Energy Information Administration.

———. 2001. *Annual Energy Outlook 2002*. Washington, DC: Department of Energy

Ellerman, A.D., P.L. Joskow, R. Schmalensee, J.P. Montero, and E.M. Bailey. 2000. *Markets for Clean Air: The U.S. Acid Rain Program*. Cambridge, UK: Cambridge University Press.

Elwell, Christine. 1996. "Sustainably Priced" Trade in Forest Products and Ecological Services: Some Legal Standards and Economic Instruments. In *Global Forests and International Environmental Law*, edited by Canadian Council on International Law. Boston: Kluwer, 193–238.

EOP (Executive Office of the President). 2002. *Climate Change Policy Book*. Washington, DC: Executive Office of the President.

Fankhauser, Samuel, Richard S. J. Tol, and David W. Pearce. 1998. Extensions and Alternatives to Climate Change Impact Valuation: On the Critique of IPCC Working Group III's Impact Estimates. *Environment and Development Economics* 3(Part 1): 59–81.

FAO (United Nations Food and Agriculture Organization). 1991. *FAO Yearbook: Forest Products, 1978–89*. Rome: FAO.

————. 1999. *State of the World's Forests, 1999.* Rome: FAO.

Fischer, Carolyn. 2001. Climate Change Policy Choices and Technical Innovation. In *Climate Change Economics and Policy,* edited by Michael A. Toman. Washington, DC: Resources for the Future.

Fischer, Carolyn, Suzi Kerr, and Michael A. Toman. 2001. Using Emissions Trading to Regulate National Greenhouse Gas Emissions. In *Climate Change Economics and Policy,* edited by Michael A. Toman. Washington, DC: Resources for the Future.

Fischer, Carolyn, and Michael A. Toman. 2001. Environmentally and Economically Damaging Subsidies: Concepts and Illustrations. In *Climate Change Economics and Policy,* edited by Michael A. Toman. Washington, DC: Resources for the Future.

Gaskins, D.W., and J.P. Weyant. 1993. Model Comparisons of the Costs of Reducing CO_2 Emissions. *American Economic Review* 83(2): 318–23.

Geller, H., S. Bernow, and W. Dougherty. 1999. *Meeting America's Kyoto Protocol Target: Policies and Impacts.* Washington, DC: American Council for an Energy-Efficient Economy.

Gould, K., A. Schaiberg, and A. Weinberg. 1997. *Local Environmental Struggles. Citizen Activism in the Treadmill of Production.* Cambridge, MA: Cambridge University Press.

Goulder, Lawrence H. 2001. Confronting the Adverse Industry Impacts of CO_2 Abatement Policies: What Does It Cost? In *Climate Change Economics and Policy,* edited by Michael A. Toman. Washington, DC: Resources for the Future.

Goulder, L.H., I.W.H. Parry, and R.C. Williams. 1999a. When Can Carbon Abatement Policies Increase Welfare? The Fundamental Role of Distorted Factor Markets. *Journal of Environmental Economics and Management* 37: 52–84.

Goulder, L.H., I.W.H. Parry, R.C. Williams, and Dallas Burtraw. 1999b. The Cost-Effectiveness of Alternative Instruments for Environmental Protection in a Second-Best Setting. *Journal of Public Economics* 72: 329–60.

Grunwald, Michael. 2000. Agency Says Engineers Likely Broke Rules: Corps Economist's Allegations of Rigged Lock Expansion Study Forwarded to Cohen. *Washington Post,* Feb. 29, A4.

Hahn, Robert W., and Robert N. Stavins. 1999. What Has Kyoto Wrought? The Real Architecture of International Tradeable Permit Markets. Discussion paper 99-30. Washington, DC: Resources for the Future. March.

Hamilton, James, and W. Kip Viscusi. 1999. *Calculating Risks: the Spatial and Political Dimensions of Hazardous Waste Policy,* Cambridge, MA: MIT Press.

Hansen, James, Makiko Sato, Reto Ruedy, Andrew Lacis, and Valdar Oinas. 2000. Global Warming in the Twenty-First Century: An Alternative Scenario. *Proceedings of the National Academy of Sciences* 97(18): 9875–80.

Hargrave, Tim. 1998. *US Carbon Emission Trading: Description of an Upstream Approach.* Washington, DC: Center for Clean Air Policy.

Harrington, Winston, Richard D. Morgenstern, and Peter Nelson. 1999. On the Accuracy of Regulatory Cost Estimates. Discussion paper 99-18. Washington, DC: Resources for the Future.

Hausker, Karl. 1992. The Politics and Economics of Auction Design in the Market for Sulfur Dioxide Pollution. *Journal of Policy Analysis and Management* 11(4): 553–72.

Hayhoe, K., A. Jain, H. Pitcher, C. MacCracken, M. Gibbs, D. Wuebbles, R. Harvey, and D. Kruger. 1999. Costs of Multigreenhouse Gas Reduction Targets for the USA. *Science* 286: 905–6.

Hecht, Susannah. 1993. The Logic of Livestock and Deforestation in Amazonia. *Bioscience* 43: 687–95.

Hendee, William. 1996. Modeling Risk at Low Levels of Exposure. In *Risks, Costs, and Lives Saved: Getting Better Results from Regulation,* edited by Robert W. Hahn. Washington, DC: Oxford University Press and AEI Press.

Holly, Chris. 2003. Allowance Plan Scrapped: Bush Clean Air Plan Faltering in Senate. *Energy Daily,* May 27.

Holt, D., and C. Bielema. 1996. Competing for Space on the Education Shelf. *Choices* (4th Quarter): 11–15.

Howarth, Richard H. 1998. An Overlapping Generation Model of Climate-Economy Interactions. *Scandinavian Journal of Economics* 100(3): 575–91.

Humphreys, David. 1996. *Forest Politics.* London: Earthscan.

IAT (Interagency Analytical Team). 1997. *Economic Effects of Global Climate Change Policies.* June draft. Washington, DC.

Interagency Task Force. 1993. *Forests for the Future: Launching Initial Partnerships.* Washington, DC. January 15.

IPCC (Intergovernmental Panel on Climate Change). 1990. *Climate Change: The IPCC Response Strategies.* Washington, DC: Island Press.

———. 1998. *The Regional Impacts of Climate Change: An Assessment of Vulnerability.* New York: Cambridge University Press.

———. 2001a. *Climate Change 2001: The Scientific Basis. Contribution of Working Group I to the Third Assessment Report of the Intergovernmental Panel on Climate Change.* Edited by J.T. Houghton, Y. Ding, D.J. Griggs, M. Noguer, P.J. van der Linden, X. Dai, K. Maskell, and C.A. Johnson. Cambridge, UK, and New York: Cambridge University Press.

———. 2001b. *Climate Change 2001: Impacts, Adaptation, and Vulnerability. Contribution of Working Group II to the Third Assessment Report of the Intergovernmental Panel on Climate Change.* Edited by J.J. McCarthy, O. F. Canziani, N. A. Leary, D.J. Dokken, and Kasey S. White. Cambridge, UK, and New York: Cambridge University Press.

———. 2001c. *Climate Change 2001: Mitigation. Contribution of Working Group III to the Third Assessment Report of the Intergovernmental Panel on Climate Change.*

Edited by B. Metz, O. Davidson, R. Swart, and J. Pan. Cambridge, UK, and New York: Cambridge University Press.

IWG (Interlaboratory Working Group). 1997. *Scenarios of U.S. Carbon Reductions: Potential Impacts of Energy Technologies by 2010 and Beyond.* Report LBNL-40533 and ORNL-444. Berkeley, CA, and Oak Ridge, TN: Lawrence Berkeley National Laboratory and Oak Ridge National Laboratory.

Jacoby, Henry D. 1999. The Uses and Misuses of Technology Development as a Component of Climate Policy. In *Climate Change Policy: Practical Strategies to Promote Economic Growth and Environmental Quality,* edited by C.E. Walker, M.A. Bloomfield, and M. Thorning. Washington, DC: American Council for Capital Formation, 151–69.

Jaffe, Adam B., Richard G. Newell, and Robert N. Stavins. 2001. Energy-Efficient Technologies and Climate Change Policies: Issues and Evidence. In *Climate Change Economics and Policy,* edited by Michael A. Toman. Washington, DC: Resources for the Future.

Johnson, Kirk. 2003. 3 States Sue EPA to Regulate Emissions of Carbon Dioxide. *New York Times,* June 5.

Kemmis, D. 2001. *This Sovereign Land.* Washington, DC: Island Press.

Kennedy, James. 2000. Risk of High Court Rejection of Standards Not a Threat to Future Rules, Browner Says. *Daily Report for Executives* 20, Jan 31, A18

Kete, Nancy. The Politics of Markets: The Acid Rain Control Policy in the 1990 Clean Air Act Amendments. Ph.D. diss., Johns Hopkins University, 1976.

Kite, M., E. Harris, and M. Thone. 1998. Visibility: A Critique of the National Programs; A Review of the Impacts in Southwest Wyoming. *Land and Water Law Review* 33(1998): 3–32.

Kolstad, Charles D., and Michael A. Toman. 2004. The Economics of Climate Policy. In *Handbook of Environmental Economics,* Volume 3, edited by K.-G. Maler and J. R. Vincent. Amsterdam: Elsevier.

Kopp, Raymond J., Richard Morgenstern, William Pizer, and Michael Toman. 1999. A Proposal for Credible Early Action in U.S. Climate Policy. Washington, DC: Resources for the Future. www.weathervane.rff.org/features/feature060.html (accessed June 3, 2004).

Kruger, Joseph, and Pizer, William A. 2004. *The EU Emissions Trading Directive: Opportunities and Potential Pitfalls.* Washington, DC: Resources for the Future.

Krugman, P. 1997. *The Age of Diminishing Expectations.* 3rd ed. Cambridge, MA: MIT Press.

———. 1998. Two Cheers for Formalism. *Economic Journal* 108(November): 1829–36.

Kuhn, Thomas. 1962. *The Structure of Scientific Revolutions.* Chicago: University of Chicago Press.

Lankton, David, Billy Pizer, Karen Palmer, and Dallas Burtraw. 2003. Legislative Comparison of Multipollutant Proposals S. 366, S. 485, and S. 843. Washington, DC: Resources for the Future. http://www.rff.org/multipollutant (accessed May 22, 2003).

Lashof, D., and D. Hawkins. 2000. *Less than Meets the Eye: An Analysis of Greenhouse Gas Emission Reductions Reported by Electric Utilities.* Washington, DC: Natural Resources Defense Council.

Lew, Jacob. 2000. Guidelines to Standardize Measures of Costs and Benefits and the Format of Accounting Statements. Memo from Jacob Lew, OMB director, to heads of departments and agencies. http://www.whitehouse.gov/omb/memoranda/m00-08.pdf.

Lexington. 2002. The Genie in the Wings. *Economist,* June 6.

López, Ramón. 2001. Including Developing Countries in Global Efforts for Greenhouse Gas Reduction. In *Climate Change Economics and Policy,* edited by Michael A. Toman. Washington, DC: Resources for the Future.

Lutter, Randall. 1999a. The Role of Economic Analysis in Regulatory Reform. *Regulation* 22(2): 38–46.

———. 1999b. Is EPA's Ozone Standard Feasible? AEI-Brookings Joint Center for Regulatory Studies, Regulatory Analysis 99-6. http://www.aei.brookings.org.

———. 2000. Developing Countries' Greenhouse Emissions: Uncertainty and Implications for Participation in the Kyoto Protocol. *Energy Journal* 21(4): 93–120.

———. 2003. Agencies' Regulatory Analyses Should Be Subject to Genuinely Independent Peer-Review. Testimony before the Water Resources and Environment Subcommittee of the House Transportation and Infrastructure Committee, Hearing on Independent Peer-Review of Scientific, Technical and Economic Products that Support Agency Decision-Making. http://aei.brookings.org/publications/abstract.php?pid=315 (accessed March 5, 2003).

Lutter, Randall, and Dallas Burtraw. Forthcoming. Clean Air for Less: Exploiting Tradeoffs between Different Air Pollutants. *Fordham Law Review.*

Lutter, Randall, and Howard Gruenspect. 2001. Assessing Benefits of Ground-Level Ozone: What Role for Science in Setting National Air Quality Standards? *Tulane Environmental Law Journal* 15(1): 85–96.

Lutter, Randall, and Elizabeth Irwin. 2002. Mercury in the Environment, a Volatile Problem. *Environment* 44(9): 24–40.

Lutter, Randall, and Christopher Wolz. 1997. UV-B Screening by Tropospheric Ozone: Implications for the National Ambient Air Quality Standard. *Environmental Science & Technology* 31: 142a–46a.

MacCracken, C. et al. 1999. The Economics of the Kyoto Protocol. *Energy Journal: Special Issue on the Kyoto Protocol,* 25–71.

Manne, A., and R. Richels. 1992. *Buying Greenhouse Insurance: The Economic Costs of Carbon Dioxide Emissions Limits.* Cambridge, MA: MIT Press.

———. 1999. The Kyoto Protocol: A Cost-Effective Strategy for Meeting Environmental Objectives? *Energy Journal: Special Issue on the Kyoto Protocol,* 1–23.

————. 2000. A Multi-Gas Approach to Climate Policy—with and without GWPs. Paper presented at EMF-19 Workshop. March 22–23, 2000, Washington, DC.

Mendelsohn, Robert. 1999. *The Greening of Global Warming*. Washington, DC: AEI Press.

Montero, Juan-Pablo. 1999. Voluntary Compliance with Market-Based Environmental Policy: Evidence from the U.S. Acid Rain Program. *Journal of Political Economy* 107(5): 998–1033.

Morgenstern, R., A. Krupnick, and X. Zhang. 2002. The Ancillary Carbon Benefits of SO_2 Reductions from a Small-Boiler Policy in Taiyuan, PRC. Discussion paper 02-54. http://www.rff.org/rff/Documents/RFF-DP-02-54.pdf (accessed September 2002).

Morita, Tsuneyuki, and John Robinson. 2001. Greenhouse Gas Emission Mitigation Scenarios and Implications. In *Climate Change 2001: Mitigation, Contribution of Working Group III to the Third Assessment Report of the Intergovernmental Panel on Climate Change*, edited by B. Metz, O. Davidson, R. Swart and J. Pan. Cambridge, UK, and New York: Cambridge University Press.

Newell, R., and W. Pizer. 2001. *Discounting the Benefits of Climate Change Mitigation: How Much Do Uncertain Rates Increase Valuation?* Report prepared for the Pew Center on Global Climate Change, Arlington, VA. December.

————. 2003. Discounting the Distant Future: How Much Do Uncertain Rates Increase Valuations? *Journal of Environmental Economics and Management* 46: 52–71.

NMA (National Mining Association). 2002. *Issues in Brief: The Clear Skies Initiative*. Washington, DC: NMA.

Nordhaus, William D. 1994. *Managing the Global Commons: The Economics of Climate Change*. Cambridge, MA: MIT Press.

————. 2001. Global Warming Economics. *Science* 294: 1283–84.

————. 2002. After Kyoto: Alternative Mechanisms to Control Global Warming. Paper presented at joint session of American Economic Association and Association of Environmental and Resource Economists. January 2, Atlanta, GA. New Haven, CT: Yale University Department of Economics.

Nordhaus, W.D., and J.G. Boyer. 1999. Requiem for Kyoto: An Economic Analysis of the Kyoto Protocol. *Energy Journal: Special Issue on the Kyoto Protocol*, 93–130.

Nordhaus, W.D., and Z. Yang. 1996. A Regional Dynamic General-Equilibrium Model of Alternative Climate-Change Strategies. *American Economic Review* 86(4): 741–56.

Novak, M. 1997. Global Climate Change Policies: The Impact on Economic Growth, U.S. Consumers, and Environmental Quality. Special Report. Washington, DC: American Council for Capital Formation Center for Policy Research. http://www.accf.org/Novak1097.htm (accessed October 1997).

————. 1998. The Kyoto Protocol: Outlook for Buenos Aires and Beyond. Testimony before the Subcommittee on Energy and Power, U.S. House of Representatives. October 6.

Oates, Wallace E., and Robert M. Schwab. 1996. The Theory of Regulatory Federalism: The Case of Environmental Management. In *The Economics of Environmental Regulation*, edited by W.E. Oates. Cheltenham, UK: Elgar.

OECD (Organization for Economic Cooperation and Development). 1995. *The OECD Reference Checklist for Regulatory Decision-Making and Background Note.* OCDE/GD(95)95. Paris: OECD.

Olson, Mancur. 1971. *The Logic of Collective Action.* 2nd ed. Cambridge, MA: Harvard University Press.

OMB (Office of Management and Budget). 2000. *Report to Congress on the Costs and Benefits of Regulation.* Washington, DC: OMB.

————. 2002. *Report to Congress on Federal Climate Change Expenditures.* Washington, DC: Office of Management and Budget.

O'Toole, Randal. 1988. *Reforming the Forest Service.* Washington, DC: Island Press.

Paltsev, S., J.M. Reilly, H.D. Jacoby, A.D. Ellerman, and K.H. Tay . 2003. *Emissions Trading to Reduce Greenhouse Gas Emissions in the United States: The McCain–Lieberman Proposal.* Report 97. Cambridge, MA: MIT Joint Center on the Policy and Science of Global Change.

Passell, P. 1998. Regulators Beware. *New York Times,* July 30.

Pearce, D.W., W.R. Cline, A.N. Achanta, S. Fankhauser, R.K. Pachauri, R.S.J. Tol, and P. Vellinga. 1995. The Social Costs of Climate Change: Greenhouse Damage and the Benefits of Control. In *Climate Change 1995: Economic and Social Dimensions of Climate Change. Contributions of Working Group III to the Second Assessment Report of the Intergovernmental Panel on Climate Change,* edited by J.P. Bruce, H. Lee, and E.F. Haites. Cambridge, UK, and New York: Cambridge University Press.

Pew Center on Global Climate Change. 2002. *Greenhouse Gas Reporting and Disclosure: Key Elements of a Prospective U.S. Program.* Washington, DC: Pew Center on Global Climate Change.

Philibert, C. and J. Pershing. 2001. Considering the Options: Climate Targets for All Countries. *Climate Policy.*

Pielke, Roger A. Jr. 1998. Rethinking the Role of Adaptation in Climate Policy. *Global Environmental Change* 8(2): 159–70.

Pizer, William A. 2001. Choosing Price or Quantity Controls for Greenhouse Gases. In *Climate Change Economics and Policy,* edited by Michael A. Toman. Washington, DC: Resources for the Future.

————. 2002. Combining Price and Quantity Controls to Mitigate Global Climate Change. *Journal of Public Economics* 85(3): 409–34.

————. 2003. *Intensity Targets in Perspective.* Paris: IFRI-RFF Conference on How to Make Progress Post-Kyoto.

Pizer, William, and Raymond Kopp. 2003. *Summary and Analysis of McCain–Lieberman—"Climate Stewardship Act of 2003."* Washington, DC: Resources for the Future.

Portney, P.R., I.W.H Parry, H.K. Gruenspecht, and W. Harrington. 2003. The Economics of Fuel Economy Standards. Discussion paper 03-44. Washington, DC: Resources for the Future. http://www.rff.org/rff/Documents/RFF-DP-03-44.pdf (accessed November 2003).

Postman, Neil. 1986. *Amusing Ourselves to Death: Public Discourse in the Age of Show Business.* New York: Viking.

Reilly, John. 2001. MIT Joint Program on the Science and Policy of Global Change. Snowmass Summer Workshop.

Reilly, J., R. Prinn, J. Harnisch, J. Fitzmaurice, H. Jacoby, D. Kicklighter, J. Melillo, P. Stone, A. Sokolov, and C. Wang. 1999. Multi-Gas Assessment of the Kyoto Protocol. *Nature* 401: 549–55.

Riebsame, W. (general ed.). 1997. *Atlas of the New West.* Center of the American West. New York: W.W. Norton and Company.

Robbins, J. 1994. *Last Refuge. The Environmental Showdown in the American West.* New York: Harper Collins West.

Roberts, Marc J., and Michael Spence. 1976. Effluent Charges and Licenses under Uncertainty. *Journal of Public Economics* 5(3–4): 193–208.

Rose, Carol M. 1998. The Several Futures of Property: Of Cyberspace and Folk Tales, Emission Trades and Ecosystems. *Minnesota Law Review* 83: 129–66.

———. 1999. Expanding the Choices for the Global Commons: Comparing New-fangled Tradable Allowance Schemes to Old-Fashioned Common Property Regimes. *Duke Environmental Law & Policy Forum* 10: 45–72.

Sandler, Todd. 1997. *Global Challenges.* New York: Cambridge University Press.

Schelling, Thomas C. 1995. International Discounting. *Energy Policy* 23: 395–401.

Schneider, David. 1996. Good Wood. *Scientific American* 274(June): 36–38.

Secretariat for Natural Resources and Sustainable Development, Argentine Republic. 1999. *Revision of the First National Communications.* Buenos Aires: Secretariat for Natural Resources and Sustainable Development.

Sedjo, Roger. 1992. Property Rights, Genetic Resources, and Biotechnological Change. *Journal of Law & Economics* 35: 199–213.

Shogren, Jason F. 1998a. A Political Economy in an Ecological Web. *Environmental and Resource Economics* 11(3–4): 557–70.

———. 1998b. *Benefits and Costs of the Kyoto Protocol.* Washington, DC: AEI Press.

Shogren, Jason, and Michael Toman. 2000. Climate Change Policy. In *Public Policies for Environmental Protection*, 2nd ed., edited by Paul Portney and Robert Stavins. Washington, DC: Resources for the Future.

Smith, Joel B., N. Bhatti, G.V. Menzhulin, R. Benioff, M.I. Budyko, M. Campos, B. Jallow, and F. Rijsberman (eds.). 1996. *Adapting to Climate Change: Assessments and Issues.* New York: Springer-Verlag.

Smith, Joel B., Hans-Joachim Schellnhuber, and M. Monirul Qader Mirz. 2001. Vulnerability to Climate Change and Reasons for Concern: A Synthesis. In *Climate Change 2001: Impacts, Adaptation, and Vulnerability. Contribution of Working Group II to the Third Assessment Report of the Intergovernmental Panel on Climate Change,* edited by J.J. McCarthy, O.F. Canziani, N.A. Leary, D.J. Dokken, and K.S. White. Cambridge, UK, and New York: Cambridge University Press.

Snow, D. 1997. Introduction. In *The Next West. Public Lands, Community, and Economy in the American West,* edited by J. Baden and D. Snow. Washington, DC: Island Press, 1–9.

Sowell, T. 1987. *A Conflict of Visions. Ideological Origins of Political Struggles.* New York: William Morrow and Company, 34.

Sperling, F. (ed.). 2003. *Poverty and Climate Change: Reducing the Vulnerability of the Poor through Adaptation.* Inter-agency report by the African Development Bank (AfDB), Asian Development Bank (ADB), Department for International Development (DFID, UK), Federal Ministry for Economic Cooperation and Development (BMZ, Germany), Ministry of Foreign Affairs–Development Cooperation (DGIS, The Netherlands), Organisation for Economic Cooperation and Development (OECD), United Nations Development Programme (UNDP), United Nations Environment Program (UNEP), and the World Bank. http://www.climatevarg.org.

Squitieri, R. 1998. Do We Always Overestimate Environmental Control Costs? U.S. Department of Treasury. Photocopy.

Stavins, Robert N. (ed.). 1988. *Project 88—Harnessing Market Forces to Protect Our Environment: Initiatives for the New President.* A Public Policy Study sponsored by Senator Timothy E. Wirth, Colorado, and Senator John Heinz, Pennsylvania. Washington, DC.

———. 1998. What Can We Learn from the Grand Policy Experiment? Positive and Normative Lessons from SO_2 Allowance Trading. *Journal of Economic Perspectives* 12(3): 69–88.

Stegner, Wallace. 1992. *Where the Bluebird Sings to the Lemonade Springs.* New York: Random House.

Stewart, Richard B. 1990. Privprop, Regprop, and Beyond. *Harvard Journal of Law & Public Policy* 13: 91, 93.

Stewart, Richard B., and Jonathan B. Wiener. 1992. The Comprehensive Approach to Global Climate Policy: Issues of Design and Practicality. *Arizona Journal of International and Comparative Law* 9: 83–113.

———. 2003. *Reconstructing Climate Policy: Beyond Kyoto.* Washington, DC: American Enterprise Institute Press.

Stiglitz, Joseph. 1997. Looking Out for the National Interest: The Principles of the Council of Economic Advisers. *American Economic Review (Papers & Proceedings)* 87: 109–13.

———. 1998. The Private Uses of Public Interest: Incentives and Institutions. *Journal of Economic Perspectives* (Spring): 3–22.

Stone, Christopher D. 1995. What to Do about Biodiversity: Property Rights, Public Goods, and the Earth's Biological Riches. *Southern California Law Review* 68: 577–620.

Subcommittee on Energy and Power. 1998a. *International Global Climate Change Negotiations.* Hearing before the Subcommittee on Energy and Power, Committee on Commerce, House of Representatives, July 15, 1997. Washington, DC: Government Printing Office.

———. 1998b. *The Kyoto Protocol and Its Economic Implications.* Hearing before the Subcommittee on Energy and Power, Committee on Commerce, House of Representatives, March 4, 1998. Washington, DC: Government Printing Office.

Swanson, Timothy. 1994. *The International Regulation of Extinction.* New York: New York University Press.

Texas Natural Resource Conservation Commission. 1999. Revisions to the State Implementation Plan (SIP) for the Control of Ozone Air Pollution. http://www.tnrcc.state.tx.us/oprd/rule_lib/pcfulsip.pdf.

Tolley, G., D. Kenkel, and R. Fabian. 1994. *Valuing Health for Policy: An Economic Approach.* Chicago: University of Chicago Press.

Toman, M.A. 2003. "Greening" Economic Development Activities for Greenhouse Gas Mitigation. RFF Issues Brief 03-02. http://www.rff.org/rff/Documents/RFF-IB-03-02.pdf (accessed December 2003).

Toman, M.A., B. Jemelkova, and J. Darmstadter. 2002. Climate Change and Economic Development. RFF issues brief 02-23. http://www.rff.org/rff/Documents/RFF-IB-02-23.pdf (accessed August 2003).

Toman, Michael A., Richard Morgenstern, and John Anderson. 1999. The Economics of "When" Flexibility in the Design of Greenhouse Gas Abatement Policies. *Annual Review of Energy* 24: 431–60.

UNCSD (United Nations Commission on Sustainable Development). 1997. *Report of the Ad Hoc Intergovernmental Panel on Forests on Its Fourth Session.* UN Doc. E/CN.17/1997/12. March 20.

———. 1999. *Report of the Intergovernmental Forum on Its Third Session.* May.

UNFCCC (United Nations Framework Convention on Climate Change). 1999a. *Convention on Climate Change.* UNEP/IUC/99/2. Geneva, Switzerland: Published for the Climate Change Secretariat by the UNEP's Information Unit for Conventions (IUC). http://www.unfccc.de.

———. 1999b. *The Kyoto Protocol to the Convention on Climate Change.* UNEP/IUC/99/10. France: Published by the Climate Change Secretariat with the

Support of UNEP's Information Unit for Conventions (IUC). http://www.unfccc.de.

U.S. Court of Appeals for the District of Columbia Circuit. 1999. *American Trucking Associations, Inc., et al., v. United States Environmental Protection Agency.* Argued December 17, 1998, decided May 14, 1999. No. 97-1440. http://pacer.cadc.uscourts.gov/common/opinions/199905/97-1440a.txt.

U.S. DOE (Department of Energy). 1995. EPA Docket A-95-54, IV-D-2694, app.

U.S. DOS (Department of State). 1997. *Climate Action Report. Submission of the United States of America under the United Nations Framework Convention on Climate Change.* Department of State Publication 10496. Washington, DC: Office of Global Change, Department of State.

U.S. EPA (Environmental Protection Agency). 1995. *Transcript of Proceedings. Public Meeting, Clean Air Scientific Advisory Committee CASAC Ozone Review Panel.* County Court Reporters. March 21, Washington DC.

———. 1996a. *Regulatory Impact Analysis for the Proposed Ozone National Ambient Air Quality Standard.* Washington, DC: U.S. EPA.

———. 1996b. *Review of the National Ambient Air Quality Standards for Particulate Matter: Policy Assessment of Scientific and Technical Information.* OAQPS Staff Paper. EPA-452\R-96-013. Washington, DC: U.S. EPA.

———. 1997a. The Benefits and Costs of the Clean Air Act, 1970 to 1990. Draft. Washington, DC.

———. 1997b. Improved Estimate of Non-Melanoma Skin Cancer Increases Associated with Proposed Tropospheric Ozone Reductions. Memo in OMB Docket 2060-AE-57 on the 1997 National Ambient Air Quality Standard for Ozone. May 22.

———. 1997c. *Fate and Transport of Mercury in the Environment.* Volume 3 of *Mercury Study to Congress.* Washington, DC: US EPA.

———. 1997d. *National Ambient Air Quality Standards: The Standard Review/Reevaluation Process, Fact Sheet.* Research Triangle Park, NC: U.S. EPA, Office of Air and Radiation, Office of Air Quality Planning and Standards. July 17.

———. 1997e. National Ambient Air Quality Standards for Ozone: Final Rule. *Federal Register* 62: 38855–96. http://www.epa.gov/fedrgstr/EPA-AIR/1997/July/Day-18/a18580.pdf.

———. 1997f. *Regulatory Impact Analysis for the Particulate Matter and Ozone National Ambient Air Quality Standards and Proposed Regional Haze Rules.* Research Triangle Park, NC: U.S. EPA, Office of Air Quality Planning and Standards. July 16.

———. 2001a. National Ambient Air Quality Standards for Ozone: Proposed Response to Remand. http://www.epa.gov/ttn/oarpg/t1/fr_notices/uvbnotic.pdf.

———. 2001b. National Ambient Air Quality Standards for Ozone; Proposed Response to Remand. *Federal Register* 66(220): 57268–92. http://www.epa.gov/fedrgstr/EPA-AIR/2001/November/Day-14/a27820.pdf.

———. 2002a. *Technical Addendum: Methodologies for the Benefit Analysis of the Clear Skies Act of 2003.* Washington, DC: U.S. EPA.

———. 2002b. *The Clear Skies Act: Technical Support Package.* Washington, DC: U.S. EPA.

———. 2002c. *An Evaluation of the South Coast Air Quality Management District's Regional Clean Air Incentives Market: Lessons in Environmental Markets and Innovation.* Washington, DC: U.S. EPA.

———. 2002d. *New Source Review: Report to the President.* Washington, DC: U.S. EPA.

———. 2003. National Ambient Air Quality Standards for Ozone: Final Response to Remand. *Federal Register* 68(3): 613–45.

U.S. FDA (Food and Drug Administration). 2001. An Important Message for Pregnant Women and Women of Childbearing Age Who May Become Pregnant about the Risks of Mercury in Fish. http://www.cfsan.fda.gov/~dms/admehg.html (accessed May 22, 2003).

U.S. OMB (Office of Management and Budget). 2002. Guidelines for Ensuring and Maximizing the Quality, Objectivity, Utility, and Integrity of Information Disseminated by Federal Agencies. http://www.whitehouse.gov/omb/fedreg/reproducible.html.

U.S. Supreme Court. 2001. *Whitman, Administrator of Environmental Protection Agency, et al. v. American Trucking Associations, Inc., et al.* http://www.supremecourtus.gov/opinions/00pdf/99-1257.pdf.

Viscusi, W. K. 1992. *Fatal Tradeoffs: Public and Private Responsibilities for Risk.* New York: Oxford University Press.

Warrick, Joby. 1998. Mass Extinction Underway, Majority of Biologists Say. *Washington Post,* April 21, A4.

WCF (World Commission on Forests). 1999. *Our Forests, Our Future.* New York: Cambridge University Press.

Weitzman, Martin L. 1974. Prices vs. Quantities. *Review of Economic Studies* 41(4): 477–91.

———. 2001. Gamma Discounting. *American Economic Review* 91(1): 260–71.

Weyant, J. 1997. Economics Impacts of Annex I Actions on All Countries. Paper presented at IPCC/EMF Workshop. August 18–20, Oslo, Norway.

Weyant, John P., and Jennifer Hill. 1999. Introduction and Overview. *Energy Journal: Special Issue on the Kyoto Protocol,* vii–xliv.

Whitfield, R.G. 1997. A Probabilistic Assessment of Health Risks Associated with Short-Term Exposure to Tropospheric Ozone: A Supplement. http://www.epa.gov/ttn/oarpg/t1sp.html (accessed January 1997).

Wiener, Jonathan Baert. 1999a. Global Environmental Regulation: Instrument Choice in Legal Context. *Yale Law Journal* 108(4): 677–800.

———. 1999b. On the Political Economy of Global Environmental Regulation. *Georgetown Law Journal* 87(3): 749–94.

———. 2001. Policy Design for International Greenhouse Gas Control. In *Climate Change Economics and Policy,* edited by Michael A. Toman. Washington, DC: Resources for the Future.

Wigley, Thomas M.L., Richard Richels, and James A. Edmonds. 1996. Economic and Environmental Choices in the Stabilization of Atmospheric CO_2 Concentration. *Nature* 379(6562): 240–43.

Wilson, E. 1998. *Consilience. The Unity of Knowledge.* New York: Knopf.

Winner, D.A., and G.R. Cass. 2000. Effect of Emissions Control on the Long-Term Frequency Distribution of Regional Ozone Concentrations. *Environmental Science and Technology* 34: 2612–17.

WRI (World Resources Institute). 1998. *World Resources, 1998–99.* New York: Oxford University Press.

Yellen, Janet. 1998. The Economics of the Kyoto Protocol. Statement before the Committee on Agriculture, Nutrition, and Forestry, U.S. Senate. March 5.

Yoon, Carol. 1993. Rain Forests Seen as Shaped by Human Hand. *New York Times,* July 27, C1.

INDEX

Acid rain policy
  under 1st Bush administration, 14
  allocation price variations, 43(n.32)
  allowance banking, 23
Adaptation measures in climate change policy, 73–74
AEA (*The Kyoto Protocol and the President's Policies to Address Climate Change: Administration Economic Analysis*). *See* Kyoto Protocol
Air pollution. *See* Clear Skies Initiative; Climate change policy; Ozone standards (1997)
Allocations
  for global forest conservation, 130–32, 139(n.39)
  updating vs. auctions in emissions trading, 24–27, 43(n.35)
Allowances, safety valves for prices of, 23–24
American Trucking (court case against EPA), 49, 53, 54, 55, 61
Aquinas, Thomas, and scholastic approach, 172–73
Argentina emissions commitments, 106, 107
Auctions vs. fixed allowances, 24–27

Banking of emissions allowances, 23
Benchmarks for evaluating potential emissions limits, 20–21
"Beneficiaries pay" model for forest conservation, 129–33, 139(n.40)
Benefit-cost analysis
  arguments for and against, 160–61
  badly done in EPA's ozone standards, 47–48, 57, 60–61
  of Clean Air Act by EPA, 155, 167–68
  and environmental policy, 19–22, 31, 62–63, 77–78, 98–99, 153–55
  of forest conservation, 120–24, 133
  See also Economics; Risk
Berlin Mandate, 91–92
  contradicted by Byrd-Hagel resolution, 81, 93
Biological diversity. *See* Forest conservation markets
Bonneville Power Authority (BPA) and electricity restructuring, 147–49
Border taxes and global forest conservation, 127–28
BPA. *See* Bonneville Power Authority (BPA)

197

Convention on Biological Diversity, 126
Convention on International Trade in
    Endangered Species (CITES), 125
Cost-benefit analysis. *See* Benefit-cost
    analysis
Cost-effectiveness
    in climate change policy, 69–70, 71–72,
        92, 94, 99–101, 112
    vs. efficiency, 11
    as goal of economists, 2–3
    See also Economics
Costs
    of climate change mitigation, 68–72, 90
    CCTI, 100–101
    downplayed by Clinton advisers, 77,
        78–79
    to U.S. economy, 92–94
    to U.S. under Kyoto Protocol, 72, 82–83,
        94–99, 102–5, 112, 156
    of electricity restructuring to BPA, 148–49
    of federal regulation, 154, 155–57
    must be considered in environmental
        policy, 63
    of ozone standards underestimated by
        EPA, 47–48, 56–61, 65(n.33),
        65(n.37), 65(n.39), 65(n.44)
Council of Economic Advisers (CEA)
    and climate change policy, 33, 81–83
    incentives for developing country commit-
        ments, 105, 106, 107–8
    and EPA's ozone standards, 48–50, 57
    presentation of policy decisions to stake-
        holders, 41
    role of, vii–xi, 1–2, 86–87, 119–20,
        149–50, 151
    support for market-based programs, 6–7,
        36–38, 129–30, 134–36
    and technology incentives in Clear Skies,
        30
    See also specific policies or topics; Yellen,
        Janet (CEA chair under Clinton)
Council on Environmental Quality (CEQ)
    and 2d Bush administration's climate
        change policy, 33
Courts. *See* Federal Appeals Court for the
    D.C. Circuit
Credits. *See* Emissions trading; Quantity-
    based regimes for global forest
    conservation

Cupitt, Larry (EPA scientist), skin cancer esti-
    mates, 51–52

Dams, environmental concerns in electricity
    restructuring, 146, 148
Decision making
    EPA's process inimical to economic rea-
        soning, 5
    improving in regulation, 61–63
    limits set by preconceptions, 5, 68
    scholastic vs. mystic approaches, 7–8,
        171–78
    See also Policy development debates
Deregulation
    and industrial performance, 143
    See also Electricity restructuring
Developing countries
    and climate change policy, 75, 80–81, 91,
        93, 96–97, 104–8
    and forest conservation, 123, 125–26,
        132–33, 135
DOE. *See* U.S. Department of Energy (DOE)

Economic Report of the President (2002), sup-
    port for institution building, 36
Economics
    advice and public policy, 3–6, 41, 153,
        163–65, 176–79
    successes and setbacks, ix–x, 166–71
    analysis
        in Clear Skies and climate change
            debates, 16–17, 39–41
        OMB guidelines for, 86–87, 178
        See also Kyoto Protocol
    economists as environmentalists, 9
    growth and emissions targets, 34–39,
        104–7, 110–11
    perceptions of public and policy makers,
        154–63
    as scholastic approach to knowledge,
        173–74
    See also Benefit-cost analysis; Cost-effec-
        tiveness; Costs; Council of Economic
        Advisers (CEA); Market-based policies;
        specific policies or topics
Efficiency
    in climate change policy, 34
    vs. cost-effectiveness, 11
    efficient market theory, 142–43, 148